Modeling a Ship's Ferromagnetic Signatures

Modeling a Ship's Ferromagnetic Signatures

John J. Holmes

ISBN: 978-3-031-00570-1   paperback
ISBN: 978-3-031-01698-1   ebook

DOI: 10.1007/978-3-031-01698-1

A Publication in the Springer series

SYNTHESIS LECTURES ON COMPUTATIONAL ELECTROMAGNETICS #16

Lecture #16

Series Editor: Constantine A. Balanis Arizona State University

Library of Congress Cataloging-in-Publication Data

Series ISSN: 1932-1252   print
Series ISSN: 1932-1716   electronic

First Edition
10 9 8 7 6 5 4 3 2 1

Modeling a Ship's Ferromagnetic Signatures

John J. Holmes
Naval Surface Warfare Center
West Bethesda, Maryland, USA

SYNTHESIS LECTURES ON COMPUTATIONAL ELECTROMAGNETICS #16

ABSTRACT

Ferromagnetic models of ships and submarines that predict or reproduce their magnetic signatures have found applications in the development of both offensive and defensive military systems from World War II to the present. The mathematical basis of generalized coordinate systems will be presented and demonstrated with example applications to analytic spherical and prolate spheroidal magnetic ship models. In addition, the advantages and pitfalls of using complex finite-element- and boundary-element numerical techniques to predict high-order near-field ship signatures will be discussed, followed by a short description of the design and testing of complementary physical scale models. Extrapolation of measured magnetic signatures from testing environments to threat areas using semi-empirical math models will be presented, along with an explanation of their inherent instabilities and methods for regularizing them. These magnetic ship signature modeling techniques are used today in designing optimized signature reduction systems that have a minimum impact on ships and their systems. The discussion will be closed with an important discussion of the verification and validation of magnetic models of surface ships and submarines.

KEYWORDS

Magnetic boundary-element models, Magnetic finite-element models, Magnetic scale models, Magnetic ship models, Model verification and validation, Multi-pole expansion, Naval vessel empirical models, Prolate spheroidal coordinates, Spherical coordinates

Acknowledgment

The author would like to sincerely thank R. Wingo, M. Lackey, and Dr. B. Hood for reviewing this manuscript and providing helpful comments.

To my brother Bill

Contents

CHAPTER 1

INTRODUCTION

The development of models that predict the ferromagnetic signatures of surface ships and submarines began in earnest during World War II to determine their susceptibilities to magnetically actuated mines and surveillance systems. These same models were also employed to evaluate and optimize the performance of signature reduction systems during their design phase. In addition, ferromagnetic models of ships and submarines have found applications in the development of both offensive and defensive military systems from World War II to the present [1].

Due to the lack of computational capability that existed during the early 1940s, mathematical models were limited to analytic formulations that were simple enough to be evaluated with slide rules or mechanical adding machines. These simple models were relegated to predicting the general characteristics of a ship's magnetic field at distances somewhat greater than its beam. At that time, predicting high-fidelity signatures very near a ferromagnetic hull could only be accomplished with detailed *physical scale models* (PSM). Magnetic scaling laws and model assembly procedures were developed, along with the construction of large specialized magnetic testing laboratories in which to measure the models' signatures. During the ensuing Cold War, PSM models remained as the primary method for predicting a ship or submarine's high-resolution ferromagnetic *near-field signatures* at distances much less than their length.

Advancements in mathematical ferromagnetic modeling of naval vessels mirrored those made in electronic computer technologies. Faster numerical processing led to improvements in the fidelity of analytic models by allowing the inclusion of higher order terms in the evaluation of complex series formulations. As high-power computer workstations became readily available, the use of numerical models, based on finite-element (FEM)- or boundary-element (BEM) methods, emerged as practical tools. When used properly, analytic and numerical math models, in combination with the PSM, can significantly reduce the cost, time, and risks associated with developing ferromagnetic ship signature reduction systems. These models can also be applied to performance predictions of offensive systems that exploit the magnetic field signatures of surface ships and submarines.

Math models can be separated into two main groups called *forward-* and *inverse-models*. Forward-models predict a ship's signature analytically based on solutions to Laplace's or Poisson's equation in separable coordinate systems that approximate the shape of a vessel's hull; or with FEM and BEM numerical simulations that use the detailed geometry of the ship's entire ferromagnetic structure and its material properties. Once these models are constructed, they can predict the triaxial signatures (longitudinal, athwartship, and vertical components) produced by the ship's permanent and induced magnetizations [1]. Forward-models can account for magnetic latitude and longitude, heading, role and pitch angles, and ship-sensor/weapon encounter geometries.

Once the spatial distribution of the magnetic signature is generated and converted into a temporal simulation for a chosen ship speed, the response of a magnetic mine or submarine surveillance system can be computed. Of course, an appropriate model of the threat device is also required. Ultimately, the overall susceptibility of a specific combatant, ship class, or battle group to these threats and the benefits of reducing their magnetic signatures constitute the desired output.

The inputs to inverse-models are magnetic signatures measured on full-scale vessels or a PSM, and accurate tracking data from which to reconstruct the ship–sensor encounter. The field and tracking data are combined to compute equivalent source strengths, which then serve as inputs to a forward analytic model for extrapolating the signatures to other sensor geometries and environments different from the original measurements. Models of this type are sometimes called *semiempirical* since they use measured data as input to solve for unknown parameters in a system of analytical equations. In fact, the outputs from FEM models have been used with inverse and analytic forward-models to reduce computational loads when predicting fields at locations beyond the arbitrary boundaries of the original mesh.

A major aspect of magnetic ship modeling is to accurately represent the effects of signature reduction systems on the uncompensated fields of a ship or submarine. The primary method for actively compensating magnetic signatures is with a *degaussing system*. A degaussing system is comprised of several loops of cable placed throughout the vessel, which when energized with the proper amount of current as established during system calibration, produce a magnetic flux distribution that is equal to the uncompensated signature but of opposite polarity. The superposition of the uncompensated or *undegaussed* signature with that of the *calibrated* degaussing loops results in a small net field. For various reasons, modeling degaussing loop signatures is challenging when using mathematical models and, in some respects, with a PSM as well.

The objective here is to describe and demonstrate magnetic modeling techniques that are applicable to naval vessels. Mathematical formulations and their application to ship models will be described in detail where warranted; however, complex derivations will be avoided in order to cover a wider range of topics. As will be shown later, solutions to Laplace's equation in the prolate spheroidal coordinate system have some major advantages in modeling the magnetic field signatures of ships. To this end, a generalized coordinate system will be given in Chapter 2 including scalar and vector operators. These equations will form the basis for generating solutions to Laplace's and Poisson's equations in the prolate spheroidal system, which is generally the one of choice for mathematically modeling ship signatures.

In Chapter 3, a vessel's induced longitudinal magnetization and signature components will be predicted with spherical and prolate spheroidal shell models. The modeling discussion will be interrupted with a brief explanation of the relationships between the nonstandard units of field and moments used historically in the magnetic signature reduction community and those of the *SI* system. Also, the effects of a ferromagnetic hull on the magnetic fields of a circular degaussing loop will be demonstrated using the spherical shell model. A short general discussion will be presented on the advantages and hurdles of the FEM and BEM numerical techniques when applying them to ship modeling. In addition, the design, construction, and testing of ferromagnetic PSM models of ships and submarines will be outlined. Scaling laws will be presented along with scale magnetic model construction techniques. The chapter will be closed with an outline of requirements for a laboratory magnetic testing facility, inside which the signatures of scale models can be measured.

One of the major objectives in processing magnetic signature data from either full-scale or scale model measurements is to represent the ship under test in the form of a distribution of equivalent sources or magnetization. These equivalent sources can then be used to regenerate interpolated signatures over a fine regular grid, called a *standard grid*, for the purpose of determining actuation contours against mines. In addition, near-field signatures can be extrapolated to the far-field with the same equivalent source model for the purpose of estimating the susceptibility of submarines to underwater or airborne magnetic surveillance systems (see [1] for an in-depth description of these threat systems). Various types of equivalent source forward-models will be listed in Chapter 4, along with inverse-modeling approaches for computing the source parameters from measured signatures. Stability issues of inverse-models, which exist in the presence of noise and tracking errors, will be covered in detail.

All the ship modeling techniques will be tied together in the summary, Chapter 5, with a discussion of model verification and validation. Model verification will be demonstrated by comparing the analytically computed fields of the induced longitudinal magnetization of a prolate spheroidal shell with those of a PSM measured inside a magnetic model laboratory. In addition, PSM signatures of a scale mode of a DE 52 class destroyer escort will be validated against those measured on full-scale vessels. Finally, improvements to existing magnetic ship modeling techniques will be suggested for future investigations.

REFERENCE

[1]   J. J. Holmes, *Exploitation of a Ship's Magnetic Field Signatures*, 1st ed. Denver, CO: Morgan & Claypool, 2006.

CHAPTER 2

Basic Equations

2.1 BASIC EQUATIONS OF ELECTROMAGNETICS

All electromagnetic theory and modeling originate with Maxwell's equations. In the *SI* system, Maxwell's equations are given in differential form by

$$\nabla \times \vec{E} = -\frac{\partial \vec{B}}{\partial t} \tag{2.1}$$

$$\nabla \times \vec{H} = \vec{J} + \frac{\partial \vec{D}}{\partial t} \tag{2.2}$$

$$\nabla \cdot \vec{B} = 0 \tag{2.3}$$

$$\nabla \cdot \vec{D} = \rho \tag{2.4}$$

where $\vec{E}$ is the electric field intensity with units of volts/meter (V/m), $\vec{B}$ is the magnetic flux density in webers/meter^2 and equals 1 tesla (T), $\vec{H}$ is the magnetic field intensity in amperes/meter (A/m), $\vec{J}$ is the electric current density with units of amperes/meter^2 (A/m^2), $\vec{D}$ is the electric displacement in coulomb/meter^2 (C/m^2), ρ is the free charge density in coulombs (C), and t is the time in seconds (s). (See Table 2.1 for a more complete set of *SI* units.) The ferromagnetic fields of ships change slowly so that they may be considered static in nature. Therefore, the time dependence in Maxwell's equations can be ignored as well as induced electric fields. Under these conditions, (2.1)–(2.4) reduce to their *magnetostatic* form

$$\nabla \times \vec{H} = \vec{J} \tag{2.5}$$

$$\nabla \cdot \vec{B} = 0. \tag{2.6}$$

For the ferromagnetic ship modeling problem, all magnetic sources will be either on or inside a vessel's hull. Therefore, the space surrounding a ship or submarine (air, seawater, and sea bottom) will be considered source free and having a magnetic permeability equal to free space ($\mu_0 = 4\pi \times 10^{-7}$ H/m), with $\vec{B} = \mu_0 \vec{H}$.

TABLE 2.1: International System of Units (*SI*)

PHYSICAL PROPERTY	SI UNITS
Mass (m)	kg
Length (l)	m
Time (t)	s
Force (F)	N (kg-m/s^2)
Pressure (Pa)	Pa (N/m^2)
Energy, work (W)	J (N-m)
Power (P)	W (J/s)
Electric charge (q)	C (A-s)
Current (I)	A
Electric field intensity (E)	V/m (N/C)
Electric potential (V)	V (J/C)
Capacitance (C)	F (C/V)
Electric dipole moment (p)	A-m
Permittivity (ε)	F/m (C^2/N-m^2)
Electric displacement (D)	C/m^2
Electrical resistance (R)	ohm (V/A)
Electrical conductance (G)	mho (A/V)
Electrical resistivity (ρ)	m-ohm
Electrical conductivity (σ)	S/m (mho/m)
Electric current density (J)	A/m^2
Inductance (L)	H (Wb/A)
Magnetic flux density (B)	T (Wb/m^2)
Magnetic flux (Φ)	Wb (V-s)
Permeability (μ)	H/m (N/A^2)
Magnetic vector potential (A)	Wb/m
Magnetic dipole moment (M)	A-m^2
Magnetic intensity (H)	A/m

Since the volume outside the hull can be taken as current free ($\nabla \times \vec{H} = 0$), and also inside when the degaussing system is off, the magnetic field in these regions may be represented by the gradient in a scalar potential (Φ_m). Placing $\vec{B} = -\mu_0 \nabla \Phi_m$ in (2.6) results in Laplace's equation

$$\nabla^2 \Phi_m = 0 \tag{2.7}$$

where $\nabla \times \nabla \Phi_m = 0$. With electric current confined to discrete conductors inside the vessel, the static magnetic boundary conditions on a ship's ferromagnetic structure are given by

$$(\vec{H_2} - \vec{H_1}) \times \hat{n} = \vec{J_s} \tag{2.8}$$

$$(\vec{B_2} - \vec{B_1}) \cdot \hat{n} = 0 \tag{2.9}$$

where $\vec{H_2}$, $\vec{H_1}$ and $\vec{B_2}$, $\vec{B_1}$ are the field intensity and flux density on each side of the boundary, respectively; $\vec{J_s}$ is the surface current density at the interface, and $\hat{n}$ is a unit vector normal to the interface and pointing from medium 1 into 2. Applying these boundary conditions produces unique solutions to Laplace's and Poisson's equations.

The current flowing inside a degaussing coil follows discrete electrical paths along well-defined vectors. Therefore, the magnetic field produced by closed electric circuits must be represented by a magnetic vector potential $(\vec{A})$. Placing $\vec{B} = \nabla \times \vec{A}$ in (2.5) and applying a vector identity results in Poisson's equation given by

$$\nabla^2 \vec{A} = \mu_0 \vec{J} \tag{2.10}$$

where $\nabla \cdot \vec{A} = 0$. The general solution to (2.10) is given by

$$\vec{A}(r) = \frac{\mu_0}{4\pi} \int_{V'} \frac{\vec{J}(r')}{|\vec{r} - \vec{r'}|} \, dv' \tag{2.11}$$

where $|\vec{r} - \vec{r'}|$ is the distance from a point on the current source to the field observation point, and (2.11) is integrated over the entire volume V' that encloses the source. The term $1/|\vec{r} - \vec{r'}|$ is Green's function, and should be expressed in a form that accommodates the analytic shape of the problem's boundaries, which is typically the hull of the ship or submarine being modeled.

2.2   COORDINATE SYSTEMS

It is always easier to solve (2.7) and (2.10) when using a coordinate system whose shape is similar to that of the body being modeled. In this section, the mathematics needed to solve these equations in the standard rectangular, cylindrical, and spherical coordinate systems will be discussed, along with the prolate spheroidal system (PSS). First, the differential vector operators in generalized curvilinear coordinates will be given from which the others can be derived.

The most basic vector operator is the gradient of a scalar potential. The generalized curvilinear coordinates are given by (q_1, q_2, q_3). (The expressions and notation will follow that found in [1].) The gradient of a scalar potential Φ is given by

$$\nabla \Phi (q_1, q_2, q_3) = \hat{e}_1 \frac{1}{h_1} \frac{\partial \Phi}{\partial q_1} + \hat{e}_2 \frac{1}{h_2} \frac{\partial \Phi}{\partial q_2} + \hat{e}_3 \frac{1}{h_3} \frac{\partial \Phi}{\partial q_3} \tag{2.12}$$

while the differential length ds_i, area da_{ij}, and volume dv are expressed as

$$ds_i = h_i\, dq_i \tag{2.13a}$$
$$da_{ij} = h_i h_j\, dq_i\, dq_j \tag{2.13b}$$
$$dv = h_1 h_2 h_3\, dq_1\, dq_2\, dq_3 \tag{2.13c}$$

where i and j run from 1 to 3 with $i \neq j$, $(\hat{e}_1, \hat{e}_2, \hat{e}_3)$ are unit vectors in the three orthogonal directions of the curvilinear coordinate system, and (h_1, h_2, h_3) are called the metrics of the system and are themselves functions of the coordinates. The metrics describe how the coordinate system curves in three dimensions.

   An example in mechanics will help to explain the importance and function of the metrics. Mechanical acceleration is the derivative of velocity, which is itself a vector. If the velocity is along a straight line, its vector does not change direction, so that the acceleration is just the time rate-of-change of the magnitude of the velocity vector. The magnitude of an object's velocity vector is called its speed. In this case, if the speed is constant, there is no acceleration. If, however, the object is traveling at a constant speed in a circle, the magnitude of the velocity vector is constant, but there is still acceleration since the object experiences centrifugal forces. This acceleration is due to the fact that the velocity vector is changing direction even though its magnitude is constant. The metrics of a curvilinear coordinate system account for effects of this type.

   From the scalar gradient follows the divergence and Laplacian operators. The divergence of a vector $\vec{H}$ in curvilinear coordinates is given by

$$\nabla \cdot \vec{H}(q_1, q_2, q_3) = \frac{1}{h_1 h_2 h_3}\left[\frac{\partial}{\partial q_1}(H_1 h_2 h_3) + \frac{\partial}{\partial q_2}(H_2 h_1 h_3) + \frac{\partial}{\partial q_3}(H_3 h_1 h_2)\right] \tag{2.14}$$

while the generalized Laplacian in curvilinear coordinates is expressed as

$$\nabla^2 \Phi(q_1, q_2, q_3) = \frac{1}{h_1 h_2 h_3}\left[\frac{\partial}{\partial q_1}\left(\frac{h_2 h_3}{h_1}\frac{\partial \Phi}{\partial q_1}\right) + \frac{\partial}{\partial q_2}\left(\frac{h_1 h_3}{h_2}\frac{\partial \Phi}{\partial q_2}\right) + \frac{\partial}{\partial q_3}\left(\frac{h_1 h_2}{h_3}\frac{\partial \Phi}{\partial q_3}\right)\right]$$
$$\tag{2.15}$$

The curl of a vector is the most complex of the vector operators in curvilinear coordinates, and is computed from

$$\nabla \times \vec{H}(q_1, q_2, q_3) = \hat{e}_1 \frac{1}{h_2 h_3}\left[\frac{\partial}{\partial q_2}(h_3 H_3) - \frac{\partial}{\partial q_3}(h_2 H_2)\right] + \hat{e}_2 \frac{1}{h_1 h_3}\left[\frac{\partial}{\partial q_3}(h_1 H_1) - \frac{\partial}{\partial q_1}(h_3 H_3)\right]$$

$$+ \hat{e}_3 \frac{1}{h_1 h_2}\left[\frac{\partial}{\partial q_1}(h_2 H_2) - \frac{\partial}{\partial q_2}(h_1 H_1)\right] \tag{2.16}$$

which can be expressed in compact form as

$$\nabla \times \vec{H}(q_1, q_2, q_3) = \frac{1}{h_1 h_2 h_3} \begin{vmatrix} \hat{e}_1 h_1 & \hat{e}_2 h_2 & \hat{e}_3 h_3 \\ \dfrac{\partial}{\partial q_1} & \dfrac{\partial}{\partial q_2} & \dfrac{\partial}{\partial q_3} \\ h_1 H_1 & h_2 H_2 & h_3 H_3 \end{vmatrix} \tag{2.17}$$

where || is the determinant operator that must be expanded from the top down. A vector can be transformed from the rectangular coordinate system (RCS) to any generalized curvilinear system with [2]

$$\begin{pmatrix} H_{q_1} \\ H_{q_2} \\ H_{q_3} \end{pmatrix} = \begin{pmatrix} \dfrac{1}{h_{q_1}} \dfrac{\partial x}{\partial q_1} & \dfrac{1}{h_{q_1}} \dfrac{\partial y}{\partial q_1} & \dfrac{1}{h_{q_1}} \dfrac{\partial z}{\partial q_1} \\ \dfrac{1}{h_{q_2}} \dfrac{\partial x}{\partial q_2} & \dfrac{1}{h_{q_2}} \dfrac{\partial y}{\partial q_2} & \dfrac{1}{h_{q_2}} \dfrac{\partial z}{\partial q_2} \\ \dfrac{1}{h_{q_3}} \dfrac{\partial x}{\partial q_3} & \dfrac{1}{h_{q_3}} \dfrac{\partial y}{\partial q_3} & \dfrac{1}{h_{q_3}} \dfrac{\partial z}{\partial q_3} \end{pmatrix} \begin{pmatrix} H_x \\ H_y \\ H_z \end{pmatrix} \tag{2.18}$$

while the inverse transform is the transpose of the square matrix in (2.18) when the system is orthogonal. Therefore, the inverse vector transform from curvilinear to RCS is given by

$$\begin{pmatrix} H_x \\ H_y \\ H_z \end{pmatrix} = \begin{pmatrix} \dfrac{1}{h_{q_1}} \dfrac{\partial x}{\partial q_1} & \dfrac{1}{h_{q_2}} \dfrac{\partial x}{\partial q_2} & \dfrac{1}{h_{q_3}} \dfrac{\partial x}{\partial q_3} \\ \dfrac{1}{h_{q_1}} \dfrac{\partial y}{\partial q_1} & \dfrac{1}{h_{q_2}} \dfrac{\partial y}{\partial q_2} & \dfrac{1}{h_{q_3}} \dfrac{\partial y}{\partial q_3} \\ \dfrac{1}{h_{q_1}} \dfrac{\partial z}{\partial q_1} & \dfrac{1}{h_{q_2}} \dfrac{\partial z}{\partial q_2} & \dfrac{1}{h_{q_3}} \dfrac{\partial z}{\partial q_3} \end{pmatrix} \begin{pmatrix} H_{q_1} \\ H_{q_2} \\ H_{q_3} \end{pmatrix} \tag{2.19}$$

Using (2.12)–(2.19) and the appropriate coordinate transformations, the solutions to (2.7) and (2.10) can be computed in any orthogonal system. For reference, the vector operators for the rectangular, cylindrical, and spherical coordinate systems are given in Appendix I; in addition to scalar and vector transformations between the rectangular, the cylindrical, and spherical systems.

2.3   PROLATE SPHEROIDAL COORDINATE SYSTEM

Since surface ships and submarines are generally longer than they are wide, the best choice of the coordinate system in which to model their magnetic fields in three dimensions is the PSS.

In this chapter, the coordinate and vector transformations between the RCS and the PSS will be provided; along with its gradient, divergence, Laplacian, curl operator, and Green's function. The PSS will then be applied to ship signature modeling in Chapter 3.

Traditionally, magnetic ship modelers have taken the x coordinate as being aligned with the longitudinal (long axis) of a vessel. However, many mathematical texts use the z-axis as the long axis in the PSS. Since this is an introduction to modeling magnetic signatures, the latter orientation of the coordinate system will be selected since it is more likely to be familiar to the reader.

Although slightly more involved, the derivation of coordinate transformations and vector operators for the PSS is identical to the cylindrical and spherical. A three-dimensional drawing of the PSS with coordinates $(\xi,\ \eta,\ \varphi)$ is shown in Fig. 2.1a, along with a two-dimensional view of a cut along the $y-z$ plane (Fig. 2.1b). The range of the ξ coordinate is from 1 to ∞, and, when held constant, produces ellipsoidal surfaces. Hyperboloid surfaces are generated when the η coordinate, which runs from -1 to 1, is held constant. The φ coordinate is analogous to that of the cylindrical and spherical systems, and, when held constant, produces planes along an angle that can range from 0 to 2π.

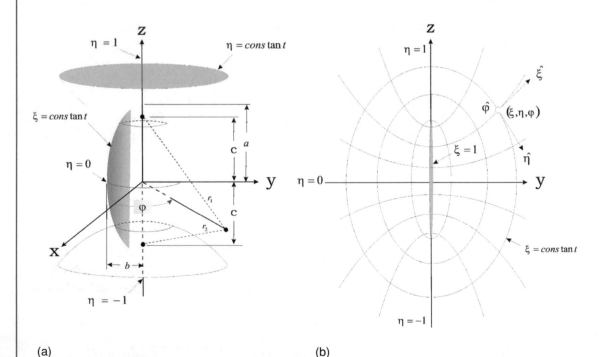

(a)   (b)

FIGURE 2.1: Prolate spheroidal coordinate system. (a) Three-dimensional view. (b) Two-dimensional view.

The coordinate transformations from the RCS to the PSS, and back, are required before proceeding to vector operators. The PSS to RCS transformations are given by [1]

$$x = c\sqrt{\left(\xi^2 - 1\right)\left(1 - \eta^2\right)}\cos\varphi \tag{2.20a}$$

$$y = c\sqrt{\left(\xi^2 - 1\right)\left(1 - \eta^2\right)}\sin\varphi \tag{2.20b}$$

$$z = c\xi\eta \tag{2.20c}$$

while the inverse transformation equations (RCS to PSS) are

$$\xi = \frac{r_2 + r_1}{2c} \tag{2.21a}$$

$$\eta = \frac{r_2 - r_1}{2c} \tag{2.21b}$$

$$\varphi = \tan^{-1}\left(\frac{y}{x}\right) \tag{2.21c}$$

where

$$r_1 = \sqrt{x^2 + y^2 + (z - c)^2}$$

$$r_2 = \sqrt{x^2 + y^2 + (z + c)^2}$$

$$c = \sqrt{a^2 - b^2}$$

and a and b are the half-length and half-width, respectively, of a spheroid with $\pm c$ as its foci. In some texts, the PSS coordinates are expressed as (u, v, φ) where $0 \leq u < \infty$, $0 \leq v \leq \pi$, and $0 \leq \varphi \leq 2\pi$, and are related to those used here by $\xi = \cosh u$, $\eta = \cos v$, and φ remains the same.

The metrics in the PSS are needed to formulate vector transformations and operators. The PSS metrics are

$$h_\xi = c\sqrt{\frac{\xi^2 - \eta^2}{\xi^2 - 1}} \tag{2.22a}$$

$$h_\eta = c\sqrt{\frac{\xi^2 - \eta^2}{1 - \eta^2}} \tag{2.22b}$$

$$h_\varphi = c\sqrt{(\xi^2 - 1)(1 - \eta^2)} \tag{2.22c}$$

Substituting the metrics of (2.22) into (2.18) and (2.19) results in the vector transformations from RCS to the PSS as given by

$$H_\xi = \xi\sqrt{\frac{1-\eta^2}{\xi^2-\eta^2}}\cos\varphi\, H_x + \xi\sqrt{\frac{1-\eta^2}{\xi^2-\eta^2}}\sin\varphi\, H_y + \eta\sqrt{\frac{\xi^2-1}{\xi^2-\eta^2}}\, H_z \qquad (2.23a)$$

$$H_\eta = -\eta\sqrt{\frac{\xi^2-1}{\xi^2-\eta^2}}\cos\varphi\, H_x - \eta\sqrt{\frac{\xi^2-1}{\xi^2-\eta^2}}\sin\varphi\, H_y + \xi\sqrt{\frac{1-\eta^2}{\xi^2-\eta^2}}\, H_z \qquad (2.23b)$$

$$H_\varphi = -\sin\varphi\, H_x + \cos\varphi\, H_y \qquad (2.23c)$$

while the inverse transforms from the PSS back to the RCS are

$$H_x = \xi\sqrt{\frac{1-\eta^2}{\xi^2-\eta^2}}\cos\varphi\, H_\xi - \eta\sqrt{\frac{\xi^2-1}{\xi^2-\eta^2}}\cos\varphi\, H_\eta - \sin\varphi\, H_\varphi \qquad (2.24a)$$

$$H_y = \xi\sqrt{\frac{1-\eta^2}{\xi^2-\eta^2}}\sin\varphi\, H_\xi - \eta\sqrt{\frac{\xi^2-1}{\xi^2-\eta^2}}\sin\varphi\, H_\eta + \cos\varphi\, H_\varphi \qquad (2.24b)$$

$$H_z = \eta\sqrt{\frac{\xi^2-1}{\xi^2-\eta^2}}\, H_\xi + \xi\sqrt{\frac{1-\eta^2}{\xi^2-\eta^2}}\, H_\eta \qquad (2.24c)$$

The coordinate and vector transforms between the two systems are orthogonal.

The complete set of scalar gradients, divergence, Laplacian, and curl in the PSS can be formulated by combining (2.22) with (2.12) through (2.16). The resulting gradient in the scalar potential Φ is given by

$$\nabla\Phi = \hat{\xi}\frac{1}{c}\sqrt{\frac{\xi^2-1}{\xi^2-\eta^2}}\frac{\partial\Phi}{\partial\xi} + \hat{\eta}\frac{1}{c}\sqrt{\frac{1-\eta^2}{\xi^2-\eta^2}}\frac{\partial\Phi}{\partial\eta} + \hat{\varphi}\frac{1}{c}\frac{1}{\sqrt{(\xi^2-1)(1-\eta^2)}}\frac{\partial\Phi}{\partial\varphi} \qquad (2.25)$$

while the divergence is

$$\nabla\cdot\vec{H} = \frac{1}{c\left(\xi^2-\eta^2\right)}\left[\frac{\partial}{\partial\xi}\left(H_\xi\sqrt{\left(\xi^2-\eta^2\right)\left(\xi^2-1\right)}\right) + \frac{\partial}{\partial\eta}\left(H_\eta\sqrt{\left(\xi^2-\eta^2\right)\left(\xi^2-1\right)}\right)\right.$$

$$\left.+\frac{\partial}{\partial\varphi}\left(H_\varphi\frac{\xi^2-\eta^2}{\sqrt{\left(\xi^2-\eta^2\right)\left(\xi^2-1\right)}}\right)\right] \qquad (2.26)$$

To complete the vector operators for the PSS, the Laplacian can be formulated as

$$\nabla^2 \Phi = \frac{1}{c^2 \left(\xi^2 - \eta^2 \right)} \left[\frac{\partial}{\partial \xi} \left(\left(\xi^2 - 1 \right) \frac{\partial \Phi}{\partial \xi} \right) + \frac{\partial}{\partial \eta} \left(\left(1 - \eta^2 \right) \frac{\partial \Phi}{\partial \eta} \right) \right.$$
$$\left. + \frac{\xi^2 - \eta^2}{\left(\xi^2 - 1 \right) \left(1 - \eta^2 \right)} \frac{\partial^2 \Phi}{\partial \varphi^2} \right] \tag{2.27}$$

while the curl is given by

$$\nabla \times \vec{H} = \frac{\hat{e}_\xi}{c} \left[\frac{1}{\sqrt{\xi^2 - \eta^2}} \frac{\partial}{\partial \eta} \left(\sqrt{1 - \eta^2}\, H_\varphi \right) - \frac{1}{\sqrt{\left(\xi^2 - 1 \right) \left(1 - \eta^2 \right)}} \frac{\partial H_\eta}{\partial \varphi} \right]$$
$$+ \frac{\hat{e}_\eta}{c} \left[\frac{1}{\sqrt{\left(\xi^2 - 1 \right) \left(1 - \eta^2 \right)}} \frac{\partial H_\xi}{\partial \varphi} - \frac{1}{\sqrt{\xi^2 - \eta^2}} \frac{\partial}{\partial \xi} \left(\sqrt{\xi^2 - 1}\, H_\varphi \right) \right] \tag{2.28}$$
$$+ \frac{\hat{e}_\varphi}{c} \left[\sqrt{\frac{\xi^2 - 1}{\xi^2 - \eta^2}} \frac{\partial}{\partial \xi} \left(\sqrt{\xi^2 - \eta^2}\, H_\eta \right) - \sqrt{\frac{1 - \eta^2}{\xi^2 - \eta^2}} \frac{\partial}{\partial \eta} \left(\sqrt{\xi^2 - \eta^2}\, H_\xi \right) \right]$$

For easy reference, the coordinate transformations and vector operators for the PSS have been included in Appendix I.

2.4   SOLUTION TO LAPLACE'S AND POISSON'S EQUATION IN PROLATE SPHEROIDAL COORDINATES

As is the case with the more popular coordinate systems, the solution to Laplace's equation in the PSS can be computed by the separation of variables method. Assuming a solution to (2.27) of the form

$$\Phi \left(\xi, \eta, \varphi \right) = \Psi \left(\xi \right) X \left(\eta \right) \Theta \left(\varphi \right) \tag{2.29}$$

where $\Psi \left(\xi \right), X \left(\eta \right), \Theta \left(\varphi \right)$ are the separation variables, and substituting (2.29) into (2.27) allows the equation to be separated into three ordinary differential equations given by

$$\frac{d^2 \Theta}{d\varphi^2} = -m^2 \Theta \tag{2.30}$$

$$\frac{d}{d\eta} \left[\left(1 - \eta^2 \right) \frac{dX}{d\eta} \right] + n \left(n + 1 \right) X - \frac{m^2}{1 - \eta^2} X = 0 \tag{2.31}$$

$$\frac{d}{d\xi} \left[\left(1 - \xi^2 \right) \frac{d\Psi}{d\xi} \right] + n \left(n + 1 \right) \Psi - \frac{m^2}{1 - \xi^2} \Psi = 0 \tag{2.32}$$

where n and m are separation constants that range over the intervals $0 \leq n < \infty$ and $0 \leq m \leq n$, respectively. The original partial differential equation (2.27) in three variables has been separated into three ordinary differential equations (2.30)–(2.32) that can be solved separately.

The solutions to the three separated ordinary differential equations are well known. Solutions to (2.30) are $\sin(m\varphi)$ and $\cos(m\varphi)$, and are harmonic functions in φ and the separation constant m. The solutions to (2.31) and (2.32), which are identical in form, are the associated Legendre functions of the first and second kind, and are symbolically represented as P_n^m and Q_n^m (degree n and order m), respectively. Series formulations, identities, and properties of P_n^m and Q_n^m are listed in Appendix V of [5].

Since (2.27) is linear, series combinations of the allowed solutions to the separated equations (2.30)–(2.32) are also valid solutions to Laplace's equation. For example, if the desired prolate spheroidal solution to (2.27) is for the space that extends from a ship's hull that might be defined by the ellipsoid $\xi = \xi_0$ outward to ∞, then the allowed general solution in this region is given by

$$\Phi(\xi, \eta, \varphi) = \sum_{n=0}^{\infty} \sum_{m=0}^{n} (A_{nm} \cos m\varphi + B_{nm} \sin m\varphi) P_n^m(\eta) Q_n^m(\xi) \qquad (2.33)$$

where A_{nm} and B_{nm} are constants to be determined through the application of boundary conditions (2.8) and (2.9), and all other terms have been defined previously.

Although both $P_n^m(\xi)$ and $Q_n^m(\xi)$ are valid mathematical solutions to (2.32), only $Q_n^m(\xi)$ is an allowed solution in this region of space since $Q_n^m(\xi) \to 0$ as $\xi \to \infty$, while $P_n^m(\xi) \to \infty$. Conversely, if the prolate spheroidal solution to Laplace's equation is desired inside the hull of a vessel $\xi \leq \xi_0$, then the allowed solution has the general form of

$$\Phi(\xi, \eta, \varphi) = \sum_{n=0}^{\infty} \sum_{m=0}^{n} (A_{nm} \cos m\varphi + B_{nm} \sin m\varphi) P_n^m(\eta) P_n^m(\xi) \qquad (2.34)$$

where $Q_n^m(\xi)$ must be omitted from the solution since $Q_n^m(\xi) \to \infty$ as $\xi \to 1$. In this case, $P_n^m(\xi)$ is allowed since it is finite at all locations inside the hull; including $\xi = 1$. However, $Q_n^m(\eta)$ must be omitted from the solutions in both (2.33) and (2.34) since $Q_n^m(\eta) \to \infty$ at $\eta = \pm 1$. The lesson here is that care must be taken when solving Laplace's equation—a function that is mathematically a valid solution may still not be allowed in the formulation since it would produce a physically unrealistic result.

Poisson's equation (2.10) is the nonhomogenous form of Laplace's. As stated at the beginning of this chapter, (2.11) is the general solution to Poisson's equation for a current source. If a current source is present on or inside a ship's hull, such as one or more degaussing coils, then solutions to Poisson's equation must be combined with the Laplace' solutions to obtain the complete answer in the volume of space that contains the sources. Conversely, the magnetic scalar potential and associated magnetic fields in source-free regions of the problem can be uniquely determined using only the solutions to Laplace's equation.

The term inside the integral of (2.11), $1/|\vec{r} - \vec{r}'|$, is called the Green's function and relates the spatial distribution of the current sources to the magnetic fields they produce. The magnetic field inside a volume V_s that contains current sources will be computed from $\vec{B}_s = -\nabla\Phi_s + \nabla \times \vec{A}_s$, while inside the source-free volumes V_f, only the magnetic scalar potential is required, $\vec{B}_f = -\nabla\Phi_f$. The unknown constants (such as the A_{nm} and B_{nm} terms) that must be evaluated for each region of the problem are determined using boundary conditions (2.8) and (2.9). A problem arises when the vector potential's integral (2.11) must be equated to the scalar potential series, such as in (2.33) or (2.34), along the entire interface between the two regions. To circumvent this mathematical difficulty, Green's functions must be expressed as series expansions in the coordinate system selected for the problem. When this is accomplished, each term in the series expansion of the vector potential can now be equated to its matching term in the series solution to Laplace's equation at each point along the boundary interfaces.

Green's functions have been formulated as series expansions for all the major coordinate systems, including the prolate spheroidal [3]. The PSS Green's function found in [4] will be used here and can be written as

$$\frac{1}{|\vec{r} - \vec{r}'|} = \frac{2}{c} \sum_{n=0}^{\infty} \sum_{m=0}^{n} \varepsilon_m (2n + 1) \left[\frac{(n - m)!}{(n + m)!}\right]^2 (\cos m\varphi_0 \cos m\varphi + \sin m\varphi_0 \sin m\varphi)$$

$$P_n^m(\eta_0) P_n^m(\eta) \begin{cases} P_n^m(\xi_0) Q_n^m(\xi); & if\ \xi > \xi_0 \\ P_n^m(\xi) Q_n^m(\xi_0); & if\ \xi < \xi_0 \end{cases} \tag{2.35}$$

where

$$\varepsilon_m = \begin{cases} 1; & if\ m = 0 \\ 2; & if\ m \neq 0 \end{cases}$$

The coordinates of the source have zero subscripts, while those without correspond to the field observation point. The reason for the two conditions in (2.35) that are dependent on the relative locations of the source and observation points is to avoid unrealistic singularities within the solution space as discussed previously.

REFERENCES

[1] G. B. Arfkin, *Mathematical Methods for Physicists*, 2nd ed. New York: Academic, 1970, pp. 72–82.

[2] P. M. Morse and H. Feshbach, *Methods of Theoretical Physics*, Part I. New York: McGraw-Hill, 1953, pp. 29–30.

[3] ——, *Methods of Theoretical Physics*, Part II. New York: McGraw-Hill, 1953, pp. 1252–1292.

[4] A. V. Kildishev and J. A. Nyenhuis, "Multipole imaging of an elongated magnetic source," *IEEE Trans. Magn.*, vol. 36, no. 5, pp. 3108–3111, Sep. 2000.

[5] C. A. Balanis, *Advanced Engineering Electromagnetics*. New York: Wiley, 1989, Appendix V.

CHAPTER 3

First-Principal Models

The input parameters to first-principal ferromagnetic ship models are the geometry of the vessel's magnetic materials, their permeability constant, and the magnitude and direction of the earth's inducing field. If degaussing coils are to be included in the simulation, then their conductor geometries and current levels are also required. First-principal modeling encompasses not only analytic and numerical simulations, but also includes magnetic physical scale models. Obviously, first-principal models fall into the general category of forward-modeling since their output is magnetic signatures, which are computed from the vessel's constitutive parameters.

First-principal models are used primarily for computing the magnetic signatures of naval ships and vessels that have yet to be constructed. The dependence of the object's induced magnetization and associated signatures on its shape and material properties is estimated with these forward models. Initial estimates of a surface ship or submarine's susceptibility to magnetic detection by mines and surveillance systems can be determined early in the design process for the new class. In addition, the impact of altering the vessel's geometry and material properties on its susceptibility to magnetic threats can be traded off against construction costs and in-water hull performance.

If susceptibility studies indicate that additional signature reduction is required for the vessel to achieve its missions, first-principal models are once again used to design and evaluate degaussing coil systems that will achieve the specified signature levels. The complexity of the degaussing coil design and its resulting reduction in the ship's susceptibility to magnetic threats are traded off against system cost, weight, volume, power, and other impacts on the hull. First-principal models are especially important if a hull's shape, magnetic properties, or level of signature reduction is dissimilar to those preexisting for which a database of actual signature measurements is available.

3.1 SPHERICAL MODEL OF A VESSEL'S HULL

The spherical shell is the simplest three-dimensional representation of an enclosed hull that can be modeled analytically. It can be used to estimate the magnetic field signatures of a diving bell, a moored mine, or as a first-order model of a ship. The formulation of a shell's magnetic

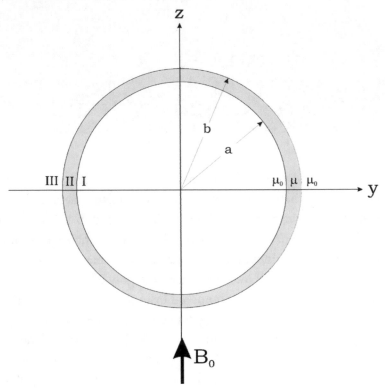

FIGURE 3.1: Spherical shell ship model to predict induced magnetic field signatures.

signature induced by a uniform external field, such as that of the earth, will serve as a simple example on how to solve Laplace's equation with two interfaces (three regions). In addition, the spherical shell's fields will be compared later to those of a prolate spheroidal shell to highlight the effects of hull geometry on its induced signatures.

Consider a vessel with a spherical hull of permeability μ placed in a magnetic field $\vec{B}_0$ that is uniform along the z-axis. As shown in Fig. 3.1, the radius to the inside and outside edges of the hull will be designated as a and b, respectively. The permeability of the space interior (Region I) and exterior (Region III) to the hull will be that of free space with permeability $\mu_0 = 4\pi \times 10^{-7}$ H/m. The hull material will occupy Region II.

Laplace's equation in spherical coordinates is given in the Appendix (I.25). Its general solution is

$$\Phi(r, \theta, \varphi) = \sum_{n=0}^{\infty} \sum_{m=0}^{n} (A_{nm} \cos m\varphi + B_{nm} \sin m\varphi) P_n^m(\cos \theta) \begin{cases} \dfrac{1}{r^{n+1}}; & if\, r \neq 0 \\ r^n; & if\, r \neq \infty \end{cases} \qquad (3.1)$$

where A_{nm} and B_{nm} are constants to be determined from boundary conditions. Since in this example the earth's magnetic field intensity $\vec{H}_0$ is taken to be uniform and aligned along the z-axis, its potential Φ_e in rectangular coordinates is $-H_0 z$, while in spherical coordinates (using (I.21c)) it is given by

$$\Phi_e = -H_0 r \cos \theta \tag{3.2}$$

Also, the applied external field and the shell's geometry are symmetrical about the z-axis, which means that (3.1) must be constant with respect to φ allowing only the $m = 0$ terms to be retained.

Internal to the spherical hull, only the lower term at the end of (3.1) may be used; otherwise, the solution would be undefined at $r = 0$. Therefore, the magnetic scalar potential in Region I (Φ_I) can be written as [1]

$$\Phi_I = \sum_{n=0}^{\infty} A_n r^n P_n (\cos \theta) \tag{3.3}$$

while inside the hull's magnetic material (Region II) $r \neq 0$ and $r \neq \infty$, so that both terms in (3.1) can be included in the expression for the magnetic potential (Φ_{II})

$$\Phi_{II} = \sum_{n=0}^{\infty} \left(B_n r^n + C_n \frac{1}{r^{n+1}} \right) P_n (\cos \theta) \tag{3.4}$$

In the exterior space, the earth's magnetic potential must be added to (3.1), where only the $1/r^{n+1}$ term is allowed in the series solution for it to remain finite as $r \rightarrow \infty$. The potential in Region III (Φ_{III}) is

$$\Phi_{III} = -H_0 r \cos \theta + \sum_{n=0}^{\infty} \frac{D_n}{r^{n+1}} P_n (\cos \theta) \tag{3.5}$$

The unknown constants A_n, B_n, C_n, and D_n will be determined from the boundary conditions along the two interfaces between the three regions.

From (3.8) and (3.9) in Chapter 2, the boundary conditions for this problem at the interfaces $r = a$ and $r = b$ state that H_θ and B_r must be continuous across them. Using (I.23), the boundary conditions can be expressed mathematically by

$$\frac{\partial \Phi_I}{\partial \theta} = \frac{\partial \Phi_{II}}{\partial \theta} \qquad \text{at } r = a \tag{3.6a}$$

$$\frac{\partial \Phi_{II}}{\partial \theta} = \frac{\partial \Phi_{III}}{\partial \theta} \qquad \text{at } r = b \tag{3.6b}$$

$$\mu_0 \frac{\partial \Phi_I}{\partial r} = \mu \frac{\partial \Phi_{II}}{\partial r} \qquad \text{at } r = a \tag{3.6c}$$

$$\mu \frac{\partial \Phi_{II}}{\partial r} = \mu_0 \frac{\partial \Phi_{III}}{\partial r} \qquad \text{at } r = b \tag{3.6d}$$

As the magnetic potential of the earth has the form of $\cos\theta$, and $P_1(\cos\theta) = \cos\theta$, the boundary conditions at $r = b$ can only be met for all θ if $n = 1$ in the series representation of the potentials in all regions. Placing (3.3)–(3.5) into (3.6a)–(3.6d) produces a system of four equations with four unknowns that are given by

$$D_1 - b^3 B_1 - C_1 = b^3 H_0 \tag{3.7a}$$
$$2D_1 + \mu' b^3 B_1 - 2\mu' C_1 = -b^3 H_0 \tag{3.7b}$$
$$a^3 B_1 + C_1 - a^3 A_1 = 0 \tag{3.7c}$$
$$\mu' a^3 B_1 - 2\mu' C_1 - a^3 A_1 = 0 \tag{3.7d}$$

where $\mu' = \mu/\mu_0$. Solving (3.7a)–(3.7d) for D_1, the only term needed to compute the magnetic signature of the spherical hull vessel in Region III, results in

$$D_1 = \left[\frac{(2\mu' + 1)(\mu' - 1)}{(2\mu' + 1)(\mu' + 2) - 2\frac{a^3}{b^3}(\mu' - 1)^2} \right] (b^3 - a^3) H_0 \tag{3.8}$$

Using (I.21c) and rewriting (3.5) in the RCS will result in the equations for the induced triaxial magnetic fields of the spherical vessel in their standard form as typically used in ship signature analysis. The exterior magnetic potential in the RCS can be simplified to

$$\Phi_{III} = -H_0 z + \frac{D_1 z}{(x^2 + y^2 + z^2)^{\frac{3}{2}}} \tag{3.9}$$

The triaxial magnetic field signatures are computed from $\vec{H} = -\nabla \Phi_m$, (3.9), and (I.3) and result in

$$H_{xIII} = \frac{3D_1 x z}{(x^2 + y^2 + z^2)^{\frac{5}{2}}} \tag{3.10}$$

$$H_{yIII} = \frac{3D_1 y z}{(x^2 + y^2 + z^2)^{\frac{5}{2}}} \tag{3.11}$$

$$H_{zIII} = H_0 + \frac{D_1(2z^2 - x^2 - y^2)}{(x^2 + y^2 + z^2)^{\frac{5}{2}}} \tag{3.12}$$

Typically, when magnetic signatures of ships are computed or measured, the earth's field is removed mathematically or with hardware filters. Ship signatures are defined as the anomalous portion of the field that is different from that of the earth. Therefore, only the last term in (3.12) represents the z component of the vessel's magnetic signature.

3.2   PROLATE SPHEROIDAL MODEL OF A VESSEL'S HULL

Typically, oceangoing naval vessels have lengths that are several times larger than their width (beam). This causes the surface ship or submarine to be magnetized differently in its *longitudinal* (long) direction than for its *athwartship* (starboard-port direction) or vertical axes. For this reason, a prolate spheroidal shell model of a naval vessel should be more representative of its induced magnetization and magnetic field signatures than the spherical shell presented in the previous section.

The geometry of a prolate spheroidal magnetic shell model of a ship is shown in Fig. 3.2, along with the coordinate system that will be used in the analysis. As was the case for the spherical shell model, the problem will be divided into three regions. The internal Region I and external Region III will have a permeability of free space, while the magnetic hull (Region II) will have a permeability of μ. The prolate spheroidal coordinates of the inner and outer interfaces of the magnetic hull will be designated as ξ_1 and ξ_2, respectively, with the earth's uniform inducing field B_0 impinging upon the hull along the z-axis.

Laplace's equation for the magnetic scalar potential in prolate spheroidal coordinates is given by (2.27) in Chapter 2. Its general solution in Regions III and I were also written in Chapter 2 as (2.33) and (2.34), respectively. Once again, the earth's inducing field is symmetric about the z-axis, as is the hull's geometry, and is not a function of φ. Therefore, to satisfy the boundary condition, only the $m = 0$ terms can be used in the series formulation of the magnetic potential in the three regions.

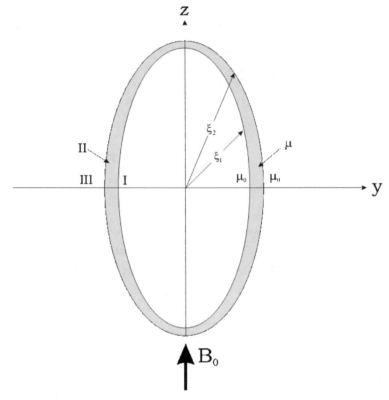

FIGURE 3.2: Prolate spheroidal shell ship model to predict induced magnetic field signatures.

The solution to Laplace's equation for the magnetic scalar potential in Regions I–III reduce to

$$\Phi_I = \sum_{n=0}^{\infty} A'_n P_n(\eta) P_n(\xi) \qquad (3.13)$$

$$\Phi_{II} = \sum_{n=0}^{\infty} \left(B'_n P_n(\xi) + C'_n Q_n(\xi) \right) P_n(\eta) \qquad (3.14)$$

$$\Phi_{III} = \sum_{n=0}^{\infty} D'_n P_n(\eta) Q_n(\xi) - H_0 c \xi \eta \qquad (3.15)$$

where P_n and Q_n are the ordinary Legendre polynomials of the first and second kind, respectively. The last term in (3.15) is the earth's magnetic scalar potential, $-H_0 z$, expressed in prolate spheroidal coordinates using (I.32c). As was the case for the spherical shell, the unknown constants A'_n, B'_n, C'_n, and D'_n will be determined from the boundary conditions along the two interfaces between the three regions.

The boundary conditions for this problem at $\xi = \xi_1$ and $\xi = \xi_2$ state that H_η and B_ξ must be continuous across the two interfaces. Due to the presence and form of the earth's magnetic potential in (3.15), the only way that the boundary conditions can be satisfied for all η is if $n = 1$. Using the identities

$$P_1(x) = x \tag{3.16}$$

$$Q_1(x) = \frac{x}{2} \ln\left(\frac{x+1}{x-1}\right) - 1 \tag{3.17}$$

and keeping only the $n = 1$ term in (3.13)–(3.15) results in

$$\Phi_I = A'_1 \eta \xi \tag{3.18}$$

$$\Phi_{II} = B'_1 \eta \xi + C'_1 \eta \left[\frac{\xi}{2} \ln\left(\frac{\xi+1}{\xi-1}\right) - 1 \right] \tag{3.19}$$

$$\Phi_{III} = D'_1 \eta \left[\frac{\xi}{2} \ln\left(\frac{\xi+1}{\xi-1}\right) - 1 \right] - H_0 c \xi \eta \tag{3.20}$$

From (I.34), the boundary conditions can be expressed mathematically as

$$\frac{\partial \Phi_I}{\partial \eta} = \frac{\partial \Phi_{II}}{\partial \eta} \qquad \text{at } \xi = \xi_1 \tag{3.21a}$$

$$\frac{\partial \Phi_{II}}{\partial \eta} = \frac{\partial \Phi_{III}}{\partial \eta} \qquad \text{at } \xi = \xi_2 \tag{3.21b}$$

$$\mu_0 \frac{\partial \Phi_I}{\partial \xi} = \mu \frac{\partial \Phi_{II}}{\partial \xi} \qquad \text{at } \xi = \xi_1 \tag{3.21c}$$

$$\mu \frac{\partial \Phi_{II}}{\partial \xi} = \mu_0 \frac{\partial \Phi_{III}}{\partial \xi} \qquad \text{at } \xi = \xi_2 \tag{3.21d}$$

Substituting (3.18)–(3.20) into (3.21a)–(3.21d) yields another system of four equations with four unknowns that are given by

$$A'_1 \xi_1 - B'_1 \xi_1 - C'_1 a_1 = 0 \tag{3.22a}$$
$$B'_1 \xi_2 + C'_1 a_2 - D'_1 a_2 = -H_0 c \xi_2 \tag{3.22b}$$
$$A'_1 - \mu' B'_1 + \mu' C'_1 a_3 = 0 \tag{3.22c}$$
$$\mu' B'_1 + \mu' C'_1 a_4 - D'_1 a_4 = -H_0 c \tag{3.22d}$$

where

$$a_1 = \frac{\xi_1}{2} \ln\left(\frac{\xi_1+1}{\xi_1-1}\right) - 1$$

$$a_2 = \frac{\xi_2}{2} \ln\left(\frac{\xi_2+1}{\xi_2-1}\right) - 1$$

$$a_3 = \frac{1}{2} \ln\left(\frac{\xi_1+1}{\xi_1-1}\right) - \frac{\xi_1}{\xi_1^2-1}$$

$$a_4 = \frac{1}{2} \ln\left(\frac{\xi_2+1}{\xi_2-1}\right) - \frac{\xi_2}{\xi_2^2-1}$$

and $\mu' = \mu/\mu_0$. Solving (3.22a)–(3.22d) for D_1' yields

$$D_1' = \frac{\xi_1\xi_2\mu'(a_3-a_4) + a_2\xi_1 - \xi_2 a_1}{\mu'^2 a_2\xi_1(a_3-a_4) + \mu'(\xi_1 a_4(2a_2 - \xi_2 a_3)) + a_4(\xi_2 a_1 - \xi_1 a_2)}\left(\mu'-1\right)c\,H_0 \quad (3.23)$$

where all terms have been defined previously. The relationship between D_1' in (3.23) and D_1 in (3.8) will be discussed later.

The magnetic signatures of naval vessels are almost always measured in the RCS, and for this reason, the triaxial induced field signatures of the prolate spheroidal shell will be computed and formulated accordingly. The triaxial magnetic fields are computed from $\vec{H} = -\nabla\Phi_m$, using (3.20) and (I.3); and, after some algebraic manipulation, are written as

$$H_{xIII} = \frac{D_1' x\eta}{r_1 r_2(\xi^2-1)} \quad (3.24)$$

$$H_{yIII} = \frac{D_1' y\eta}{r_1 r_2(\xi^2-1)} \quad (3.25)$$

$$H_{zIII} = H_0 + \frac{D_1'}{c}\left[-\frac{1}{2}\ln\left(\frac{\xi+1}{\xi-1}\right) + \frac{c^2\xi}{r_1 r_2}\right] \quad (3.26)$$

where ξ, η, c, r_1, and r_2 have all been defined in the Appendix in terms of x, y, and z. As was discussed previously for the spherical shell, only the last term in (3.26) represents the z component of the vessel's magnetic signature, which is detected as an anomaly in the earth's normally uniform field.

Formulating the magnetic field signatures of a prolate spheroidal shell that has an inducing field along its x- or y-axis is somewhat more involved than the longitudinal example given here. The mathematical expressions for the magnetic scalar potentials and moments of both a prolate and oblate spheroidal shell with an arbitrary inducing field are given in [2] with the solutions expressed in determinant form, while a complete set of algebraic equations describing the triaxial magnetic field signatures can be found in Appendix E of [3].

If the length of a prolate spheroid is set equal to its width, then (3.23)–(3.26) should predict the same signatures as the spherical shell equations, (3.8)–(3.12). Indeed, if both sets of equations are numerically evaluated, they produce identical x and z signature components shown as the inner solid curves in Figs. 3.3 and 3.4, respectively. These figures also demonstrate how the induced longitudinal magnetic signatures change with the spheroid's length. Holding the earth's field, and the simulated ship's beam and bow/stern hull thickness constant, the peaks in the vertical component (Fig. 3.3) increase in amplitude and move farther apart, keeping just inside the ends. (In spite of holding the inside radii (a_i) of the bow and stern constant, the inside radius at the middle of the prolate spheroid (b_i) must change with its length according to $b_i = \sqrt{a_i^2 - c_o^2}$, where c_o is the semifocal length of the outer surface of the hull.) However, the peak in the longitudinal component increases at first with the ship's length and midship's hull

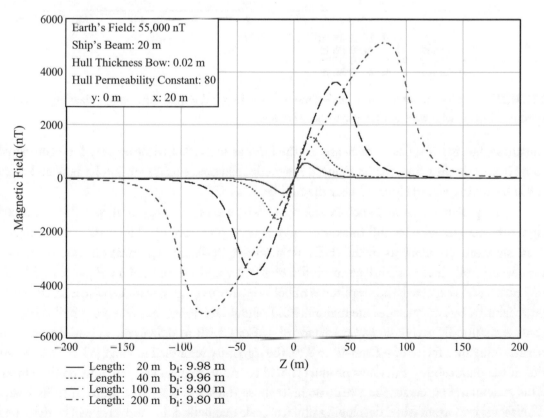

FIGURE 3.3: Vertical component signatures of the induced longitudinal magnetization of prolate spheroidal shells with different length-to-beam ratios.

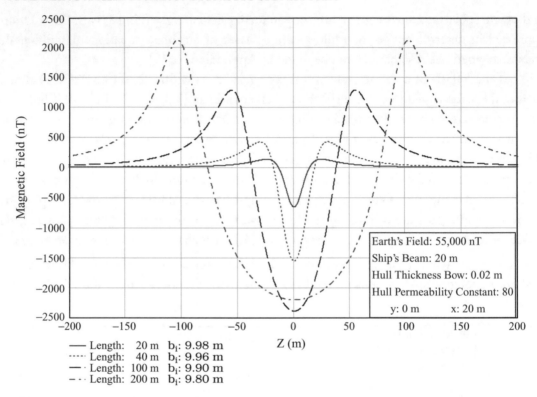

FIGURE 3.4: Horizontal component signatures of the induced longitudinal magnetization of prolate spheroidal shells with different length-to-beam ratios.

thickness, but then begins to decrease. In the limit as the vessel becomes very long compared to its beam, the flux lines for its induced magnetic field concentrate around its bow and stern while becoming extremely small near the middle.

The prolate spheroidal shell model can also be used to investigate how a ship's induced signature changes with its hull thickness and magnetic permeability. The vertical and longitudinal signature components of the shell's induced longitudinal magnetization were computed for several hull thickness and permeability constants, and are plotted in Figs. 3.5 and 3.6, respectively. As expected, the amplitudes of both signature components decrease with the hull's permeability, but their shapes are maintained. It should be noted that when the hull thickness at its ends/middle was reduced by a factor of 4, from 2/20 to 0.5/5 cm, while increasing its permeability by a factor of 4 from 80 to 320, the signature remained unchanged. This suggests that if the permeability-thickness product is held constant, then so will the induced signatures. This statement is accurate for a realistic hull whose thickness is small compared to its beam. This fact is important when developing physical scale magnetic ship models as will be discussed later.

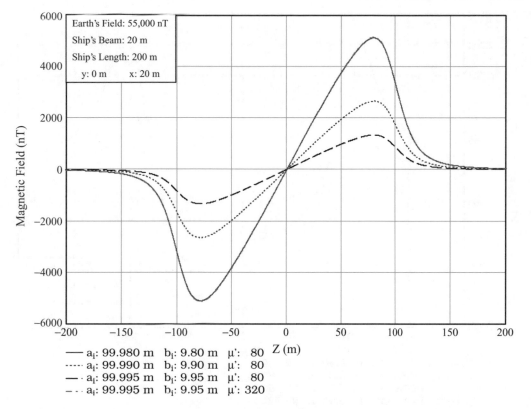

FIGURE 3.5: Vertical component signatures of the induced longitudinal magnetization of prolate spheroidal shells with different permeability-thickness products.

3.3   DIPOLE MOMENTS AND THEIR UNITS

It is tempting to think of D_1 and D_1' as the dipole moments for their respective vessel's induced magnetization. However, in the *SI* system being used here, a magnetic dipole moment is defined in terms of electric current, I, in amperes flowing in a loop of wire that encloses an area, A, in units of meters-squared. A z-directed spherical magnetic dipole $m_z = IA$ in the *SI* system with units of A-m^2 generates magnetic fields (in tesla) according to the equations

$$B_x = \frac{\mu_0 3 m_z x\, z}{4\pi (x^2 + y^2 + z^2)^{\frac{5}{2}}} \tag{3.27}$$

$$B_y = \frac{\mu_0 3 m_z y\, z}{4\pi (x^2 + y^2 + z^2)^{\frac{5}{2}}} \tag{3.28}$$

$$B_z = \frac{\mu_0 m_z (2z^2 - x^2 - y^2)}{4\pi \left(x^2 + y^2 + z^2\right)^{\frac{5}{2}}} \tag{3.29}$$

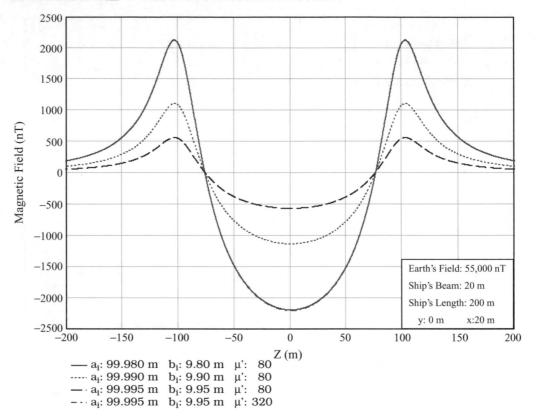

FIGURE 3.6: Horizontal component signatures of the induced longitudinal magnetization of prolate spheroidal shells with different permeability-thickness products.

Therefore, D_1 in (3.8) must be multiplied by 4π before it can be equated to the spherical magnetic dipole moment $m_z = 4\pi D_1$ and used in the *SI* system's equations for magnetic field given by (3.27)–(3.29).

It should be noted that $m_z \neq 4\pi D_1'$. The reason is that both m_z and $4\pi D_1$ represent the moment of a spherical dipole as defined in the spherical coordinate system. Conversely, $4\pi D_1'$ represents the moment of a prolate spheroidal dipole formulated in the prolate spheroidal coordinate system. They are not equal. However, as will be shown in the next chapter, there are advantages if spherical and prolate spheroidal dipole moments can somehow be combined and intermixed in hybrid-type semiempirical magnetic models of surface ships and submarines. To aid in this mixing of dipole types, conversions from one to the other are useful.

It turns out that spherical and prolate spheroidal dipole moments are mathematically related. If D_1' is replaced with $3D_1/c^2$ then from [4], (3.24)–(3.26) will produce the same far-field signatures as their respective spherical equations (3.10)–(3.12). The far-field static magnetic signature of a naval vessel has the form of a dipole at measurement distances much greater than the ship's length. For a prolate spheroid, this occurs when $x^2 + y^2 + z^2 \gg a^2$. Obviously, for distances much less than a *ship-length*, neither signature shapes nor amplitudes are similar between dipoles in the two coordinate systems.

Historically, the ship and submarine magnetic signature reduction community has worked in a mixture of the *emu* system of units for electrical quantities, and *English* units for distance and weight. As can be imagined, converting to the *SI* system has caused and is still causing confusion and errors. Since *SI* is today the accepted system of units for conducting and reporting scientific investigations, it is imperative that conversion from the "old" system (*emu-English*) to the *SI* be performed correctly and accurately.

In the old system, magnetic signatures were measured and reported in terms of the field intensity, H, which has units of gamma (γ) with the distance between the magnetic source and the field measurement point given in feet. In the *SI* system, magnetic flux density, B, is measured in units of tesla (T) or more typically in nanotesla (nT), while distance is measured in meters. As it turns out, 1 γ of field intensity in the *emu* system equals 1 nT of flux density in the *SI*. This is simple enough. However, the confusion starts when dipole or higher order moments are being converted.

In the old system, formulations for computing static magnetic field signatures from dipoles are given by (3.10)–(3.12). In this case, the dipole moment has the units of $\gamma - \text{ft}^3$. In the *SI* system, the dipole moment has units of A-m^2, with the flux density computed using (3.27)–(3.29). If the moment conversion is attempted by simply converting γ to nT and then A/m, and ft^3 are changed to m^3, the resulting dipole moment value will not be correct for use in the *SI* system. As explained earlier, the *emu* moments must also be multiplied by 4π (dimensionless) to account for the fact that *SI* moments are defined in terms of current and area. Therefore, the correct conversion of dipole moments from the old system to the *SI* is $1\,\gamma - \text{ft}^3 = 2.8317 \times 10^{-4} \text{A-m}^2$.

3.4   MATHEMATICAL MODELS OF A DEGAUSSING COIL

Magnetic field models of surface ships and submarines are the primary tools used in the development of signature reduction systems. With them, the benefits of employing low magnetic materials in the construction of all or part of a vessel can be weighed against the additional cost of using them. When a ship's magnetic material content has been reduced as much as is technically feasible or affordable and additional field reduction is still required, then active systems are designed to compensate or cancel its magnetic field down to acceptable levels. Degaussing coils have been the traditional system of choice to actively compensate the magnetic signatures of naval vessels.

Degaussing coils are multiconductor cables that are placed inside a ship to cancel the magnetic fields produced by its induced and permanent magnetization. Several separately powered and controlled cable loops are installed in each of three orthogonal planes to precisely compensate the longitudinal, athwartship, and vertical components of magnetization. Each degaussing loop included in the design adds another degree of freedom that can be controlled to increase the degree of fidelity in canceling the signature. However, the contribution of each loop in the design and the benefits of adding another to further reduce the ship's signature must be weighed against the additional cost of its construction and installation. Therefore, accurate models of degaussing coils are important not only for achieving optimum signature reduction, but to also ensure that it is achieved at minimum system cost and impact on the ship.

To maximize the efficiency of degaussing coils in generating magnetic fields, they are usually installed just inside the vessel's hull to achieve maximum area and magnetic moment. The consequence is that the offboard distance at which the signature is to be reduced (the mine threat distance) can be much less than the diameter of the coil. Therefore, degaussing coils cannot usually be modeled with point dipoles. Coil models must be used that can predict the fields from a loop of finite size. Ultimately, the models must also account for the noncircular shape of the coil as it follows the contour of the vessel's hull.

The simplest formulation of a degaussing coil is a single loop of current in free space with no permeable boundaries nearby. Analytic equations that can predict the magnetic field produced by a current loop of finite diameter can be found in [5]. Using the cylindrical coordinate system as defined in Appendix I, the radial magnetic field, B_ρ, and the longitudinal field, B_z, are given by

$$B_\rho = \frac{\mu_0 I}{2\pi} \frac{z}{\rho[(R_1 + \rho)^2 + z^2]^{\frac{1}{2}}} \left[-K(k) + \frac{R_1^2 + \rho^2 + z^2}{(R_1 - \rho)^2 + z^2} E(k) \right] \qquad (3.30)$$

$$B_z = \frac{\mu_0 I}{2\pi} \frac{1}{\rho[(R_1 + \rho)^2 + z^2]^{\frac{1}{2}}} \left[K(k) + \frac{R_1^2 - \rho^2 - z^2}{(R_1 - \rho)^2 + z^2} E(k) \right] \qquad (3.31)$$

where

$$k = \sqrt{\frac{4R_1\rho}{(R_1 + \rho)^2 + z^2}}$$

$$K(k) = \int_0^{\frac{\pi}{2}} \frac{d\vartheta}{\sqrt{1 - k^2 \sin^2 \vartheta}}$$

$$E(k) = \int_0^{\frac{\pi}{2}} \sqrt{1 - k^2 \sin^2 \vartheta} \, d\vartheta$$

and R_1 is the coil's radius, I is the coil current, $K(k)$ and $E(k)$ are the complete elliptic integrals of the first and second kind, and all other parameters have been defined previously. It should be noted that if the loop is made up of more than one conductor carrying current then I should be replaced with the total ampere-turns inside the cable. Since there is no magnetic material either inside or outside the loop, the change in signature that it produces when energized with 1 A is sometimes called the *air-core loop-effect*.

A common approach to modeling air-core loop-effects for degaussing coils of arbitrary shape is to approximate the cable runs with straight line segments. The Biot–Savart law can then be applied to each current segment to determine its field contribution to the total loop-effect (see Appendix II). Of course, care must be taken during software encoding to properly rotate and translate the magnetic fields computed for each segment into the ship's coordinate system. Although air-core loop-effects do not account for the influences of the ship's magnetic hull, experience has shown that in general the shapes do not change significantly between the two, while their amplitude differences can be estimated based on the location and orientation of the loop. Due to the ease in modeling degaussing coils with air-cores and the speed with which their fields can be computed, amplitude-adjusted air-core loop-effects are used regularly for the initial degaussing coil design. At later stages in the ship's design, more realistic but resource-intensive numerical and physical scale models are used to check the initial design and to produce final ampere-turn specifications for each loop.

Interaction of the degaussing coil's field and the ship's magnetic hull must be taken into account at some point in the design process. Analytic equations for the fields produced by a current loop of finite diameter and width located inside and outside a spherical shell were derived in [6]. However, only the exterior magnetic field signatures from an internal degaussing coil are of interest here, and will be presented without rederiving them.

The geometry of the spherical shell with an internal degaussing coil is similar to that used previously for computing its induced signature. A degaussing loop in the form of an infinitely thin current band has been added to the spherical shell of Fig. 3.1 and redrawn in Fig. 3.7. The coil has a finite diameter whose outer edge follows the contour of the circle with radius R_1. The angle that subtends the outer corners of the finite-width current band is equal to 2α. The surface current density J on the band is constant and only has a φ component.

Since the surface current density of the degaussing loop has only a φ component, then the problem is symmetric about the z-axis and can be uniquely solved using a single vector potential component, A_φ. Under this condition and using (I.26), the magnetic fields B_r and B_θ

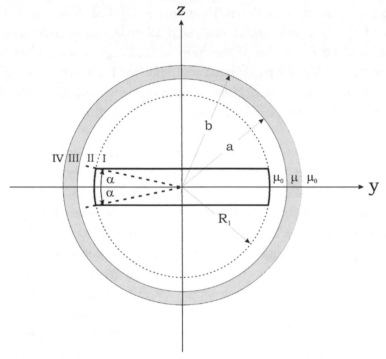

FIGURE 3.7: Model of a degaussing coil inside a spherical shell ship model.

can be expressed as

$$B_r = \frac{1}{r \sin \theta} \frac{\partial}{\partial \theta} \left(\sin \theta \, A_\varphi \right) \tag{3.32}$$

$$B_\theta = -\frac{1}{r} \frac{\partial}{\partial r} \left(r \, A_\varphi \right) \tag{3.33}$$

As shown in Fig. 3.7, the problem is broken into four regions instead of three as was done for the shell induction problem. Region I includes the spherical volume inside the boundary formed by the infinitely thin current band and Region II the space between the band and the surface of the permeable hull. Region III is the volume inside the steel hull, while Region IV includes all space outside the spherical hull. The vector potential in Region IV is the desired solution that will be used to compute the magnetic signature or *loop-effect* of the internal degaussing coil.

The vector potential solutions in the four regions are required even though Region IV is the only one of interest here. The general solutions to Poisson's equation in the four regions are given by [6]

$$A_{\varphi 1} = \sum_{p=1}^{\infty} \left(A_p r^p\right) P_p^1 \left(\cos\theta\right) \qquad (3.34a)$$

$$A_{\varphi 2} = \sum_{p=1}^{\infty} \left(B_p r^p + \frac{C_p}{r^{p+1}}\right) P_p^1 \left(\cos\theta\right) \qquad (3.34b)$$

$$A_{\varphi 3} = \sum_{p=1}^{\infty} \left(D_p r^p + \frac{E_p}{r^{p+1}}\right) P_p^1 \left(\cos\theta\right) \qquad (3.34c)$$

$$A_{\varphi 4} = \sum_{p=1}^{\infty} \left(\frac{F_p}{r^{p+1}}\right) P_p^1 \left(\cos\theta\right) \qquad (3.34d)$$

where $A_{\varphi 1}$ through $A_{\varphi 4}$ are the vector potentials in the four regions, P_p^1 is the associated Legendre function of the first kind with degree p and order 1, and A_p, B_p, C_p, D_p, E_p, and F_p are constants to be determined from the boundary conditions (2.5) and (2.6) in Chapter 2. Since the surface current density is nonzero on the sphere with radius R_1 between the angles of $\theta = \pi/2 - \alpha$ to $\theta = \pi/2 + \alpha$, it must be expanded in terms of P_p^1 in order to match boundary conditions for all θ at the interface between Regions I and II (see [6] for details of the expansion). Applying the boundary conditions at the three interfaces between the four regions results in six equations from which to solve for A_p through F_p. The system of equations was solved algebraically in [6], where F_p can be written as

$$F_p = D_p \left(b^{2p+1} + X_p\right) \qquad (3.35)$$

where

$$D_p = \frac{\mu' \left[J_p' \left(2p + 1\right) a^{-(p+2)}\right]}{\mu' \left(p + 1\right) a^{p-1} + \mu' X_p \left(p + 1\right) a^{-(p+2)} - \left(p + 1\right) a^{p-1} + p X_p a^{-(p+2)}}$$

$$X_p = -\frac{b^{2p+1} \left[\mu' + \frac{p+1}{p}\right]}{\mu' - 1}$$

$$J_p' = \frac{\mu_0 J K_p R_1^{p+2}}{2p + 1}$$

$$K_p = \frac{2p + 1}{p \left(p + 1\right)} \int_0^{\sin\alpha} P_p^1 \left(\beta\right) d\beta$$

and all other terms have been defined previously. Placing (3.35) into (3.32) and (3.33), and transforming them to the RCS with (I.22), results in expressions for the loop-effects of a single degaussing coil inside a spherical magnetic hull.

To check the equations presented in this section, the signatures computed with μ' set equal to 1 in (3.35) should be identical to those obtained with the air-core formulations of (3.30) and (3.31). To perform this verification, both sets of equations were evaluated at a sensor depth of 20 m and a degaussing loop with a radius of 9.68 m and carrying a current of 1 A. Although inconsequential to this comparison, the inner radius of the nonmagnetic hull represented by the spherical shell was kept at 9.98 m and the outer radius at 10.00 m. The vertical signatures for the air-core degaussing loop and the one inside the nonmagnetic hull are plotted as a solid and a dotted curve, respectively, in Fig. 3.8. The same was done for the longitudinal component as shown in Fig. 3.9. The solid and dotted curves in both figures are identical and lie on top of each other.

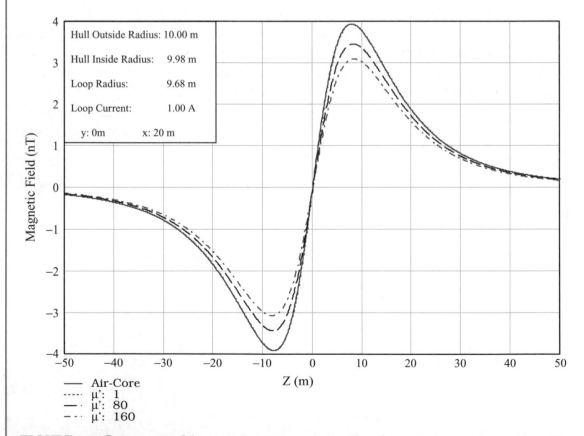

FIGURE 3.8: Comparison of the vertical component loop-effects from a horizontal degaussing coil in air to those predicted for an enclosing permeable spherical shell.

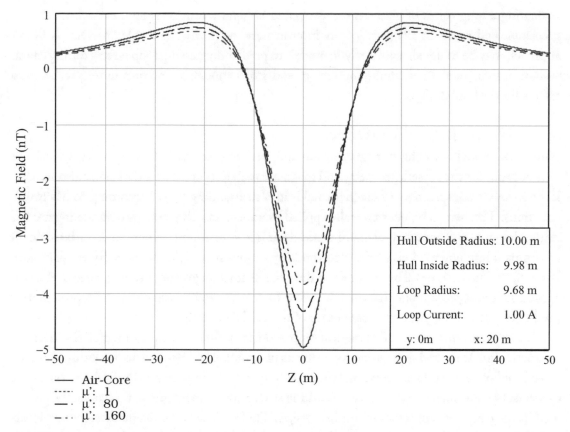

FIGURE 3.9: Comparison of the horizontal component loop-effects from a horizontal degaussing coil in air to those predicted for an enclosing permeable spherical shell.

When the permeability of the hull is allowed to increase from 1, some of the magnetic flux generated by the loop is shunted away from the exterior space surrounding the hull. As shown in Figs. 3.8 and 3.9, the coil-effect amplitude is reduced by approximately 12% for a hull thickness of 0.02 m and a permeability constant of 80, while a 20% reduction in field is predicted if the hull's permeability constant is raised to 160. The ampere-turns of the loop will have to be increased proportionally to compensate for this attenuation. A 20% increase in a degaussing system's cost, weight, and power is very undesirable, and is a topic for another time.

Analytic equations for the fields produced by a current loop of finite diameter and width located inside and outside a prolate spheroidal shell can be found in [7]. The field equations are expressed as slowly convergent alternating series of associated Legendre functions of the first and second kind. The terms in the series are products of divergent alternating sequences with those that are convergent. Although the overall formulations are convergent, issues in evaluating a long series of high-degree associated Legendre functions require high numerical precision, and result in computationally intensive and somewhat unstable answers. A better approach may be to use air-core analytic models to predict degaussing loop effects during initial system designs, and then employ numerical and PSM models to predict more precise loop effects for detailed designs.

3.5   NUMERICAL MODELS

Numerical models of a ship's magnetization and resulting magnetic field signatures must not only account for the vessel's geometry and magnetic material properties, they must also numerically simulate the spreading of magnetic fields into surrounding spaces according to Maxwell's equations. The two techniques typically applied to underwater ship signature modeling are the finite-element- and boundary-element methods. It is not the intention here to describe in detail or compare implementations of these methods by various investigators and software packages. Instead, the discussion will be confined to the general advantages and disadvantages of the two approaches, and precautions that must be taken to produce stable, accurate, and representative predictions of the magnetic field signatures of the modeled vessel.

The main advantage of numerical ship models is their ability to predict the induced magnetic signatures and degaussing loop-effects of complex hull geometries including field interactions between various internal magnetic structures and components. Preliminary estimates of a vessel's undegaussed signature and resulting susceptibility to magnetic threats can be generated in the early concept stages of a ship's design. Tradeoffs between magnetic signature levels and hull geometry, internal structure, placement of machinery and equipment items, selection of the magnetic properties of construction materials, and electric power system configurations must be made at a very early stage in the overall ship design; which, in practice, is the only time these items can be changed.

If required, numerical models can determine the signature compensation performance of various degaussing coil configurations using realistic routing of its cables. From this information, preliminary electrical drawings could be generated for the new ship class. Also, the approximate cost, weight, power, and space requirements of the proposed signature reduction concepts can be quantified early enough to avoid costly and time-consuming changes to the naval vessel's design and construction. As the overall ship design progresses, the numerical models can be updated and rerun for refined estimates of ship susceptibility and impact of signature reduction approaches.

In the past, the finite-element method (FEM) has been the preferred approach to numerically model ferromagnetic field signatures of naval vessels. This technique approximates solutions to Maxwell's equations by discretizing all space inside and outside the ship, in addition to the magnetic hull itself. Magnetic potentials and fields can be computed at each node in the discretized space (*mesh*), while interpolated solutions are generated at locations in between the nodes.

Unlike traditional applications of the magnetic FEM technique to electromechanical machinery and electric power transmission and distribution equipment, modeling the magnetization and field signatures of naval vessels has its own unique challenges. Some of the more important magnetic FEM ship modeling hurdles are:

1. The surrounding space outside the hull is infinite in nature (open-boundary problem), thus increasing modeling and computational difficulties over the better-conditioned closed-boundary simulations.

2. Depending on the geometric complexity of a vessel's magnetic hull and structure as incorporated into the model and the fidelity required of its predicted signatures, the FEM mesh size and resulting computational demand on the computer selected for these simulations can be significant.

3. High numerical accuracy is required for predicting fields that may approach 1 T in amplitude near onboard sources, but must also reproduce signatures of the order of pT at the locations of threat sensors.

4. High FEM mesh density is required to accurately account for the induced magnetization of extremely thin steel plates that run the entire length and width of the vessel.

5. High field gradients exist near degaussing coils and electric distribution cables that force high FEM mesh density.

These problems are compounded by the need to frequently repeat computations to optimize the design of signature reduction systems. As a result, there is constant pressure to reduce FEM model complexity and node count, which may seduce the modeler into taking unjustified shortcuts leading to inaccurate or nonrepresentative predictions of the vessel's signatures and susceptibilities. In the worst case, the signature reduction system may not protect the naval platform from mines or magnetic detection.

Commercially produced FEM software packages are constantly being improved to eliminate or circumvent the magnetic ship modeling difficulties listed earlier. Noteworthy solutions to these problems have been developed as reported in [8] and [9]. Improvements to FEM ship modeling can be accomplished by:

1. Using the *reduced scalar potential* for computing a ship's induced magnetization.

2. Converting the open boundary problem into a closed boundary with spatial transformations and mapping techniques.

3. Reducing the steel plate from a high mesh density volume to an infinitely thin surface element with an analytic continuation technique valid for thin plates with nontrivial permeability.

4. Eliminating the need for high mesh densities near degaussing coils through a method called the *reduced potential jump*.

The application of these FEM improvements to magnetic ship modeling can significantly reduce computation time without loss of accuracy. In-depth explanations of these special FEM modeling techniques and their performance enhancements are given in [8] and [9].

Although, in principle, the boundary-element method (BEM) can be applied to modeling magnetic ship signatures, techniques to circumvent its major drawbacks for this application have not as yet been developed. The advantage of BEM ship models is that only the boundaries of its magnetic material need to be digitized with a mesh and not the entire space inside and outside the hull. The unknown magnetic potentials or fields are computed only at the nodes on the material boundaries instead of the entire volume of space. With BEM postprocessing, the magnetic field signatures need to be computed only at the points of interest. However, because of the large number of magnetic material boundaries that exist in most ship models, some of which are very close together, the resulting computational load and numerical precision demands of the BEM approach have made it less attractive than the FEM for this application. This problem is aggravated when the nonlinear properties of a material's magnetic hysteresis must be simulated. Hybrid modeling approaches have been proposed that combine the best characteristics of the BEM and FEM for optimum performance [10].

3.6   FERROMAGNETIC PHYSICAL SCALE MODELS

Magnetic scale models of ships were first developed by British and American scientists during the early part of World War II. The models were constructed as tools in the design of degaussing coils for naval vessels to reduce their susceptibility to actuating German magnetic mines. In the United States, the first magnetic model was constructed to reproduce the signatures of the aircraft carrier USS WASP (CV-7). Although this scale model was a crude first-order approximation to the structure of the actual ship, signature comparisons between the two were close enough to justify starting a much larger modeling effort. This investigation delved into questions dealing with the requirements for model detail and design, magnetic model construction techniques, laboratory testing facilities, and the development of miniature magnetometers for making measurements close to the model. New ship classes that were slated to start construction had the highest priority since the resources needed to install permanent degaussing coils was and still is lowest when building the hull. After 1943, degaussing coil designs were becoming standardized to a point where subsequent model studies focused instead on magnetic weapons performance as opposed to ship self-protection. In total, about 60 scale magnetic models were constructed in the United States during the war, and through modifications, represented 75 different ship classes [11].

Two techniques have been used to scale magnetic models of naval vessels. The first called *thickness scaling* is a straightforward scaling of all dimensions, including the thickness of the hull, while using the same material (steel) in the construction of the scale model as to be used for the full-scale vessel. This technique is used when the model's scaled components are still large enough to be machined or cold-rolled. However, care must be taken during the model's manufacturing process so as not to significantly change the materials' resultant permeability. On occasion, an annealing treatment is applied to the completed model to compensate for these changes.

For some ship classes, the scaled hull thickness is too small to be manufactured to precise specifications without it deforming during model construction and testing. Under these conditions, a second method is employed called *permeability-thickness modeling*. When using this technique, the scaled permeability-thickness of the model's hull is maintained even though it may be constructed from materials with magnetic and mechanical properties different than those of the full-scale ship. Permeability-thickness modeling yields a mechanically stronger and less fragile model while maintaining the desired magnetic characteristics. Although an in-depth description of this scale modeling technique can be found in [12], a brief introductory overview will be presented here.

As shown in Figs. 3.5 and 3.6, the induced magnetic signatures of the prolate spheroidal shell mathematical ship model did not change when the hull thickness was reduced by a factor of 4 and its permeability was also simultaneously increased by a factor of 4. The modeled hull's permeability-thickness and its resultant induced signatures remained constant. If the scaled plate thickness and magnetic permeability constant are given by t_m and μ_m, respectively, then their product can be computed from the scaling equation

$$t_m \mu_m = \frac{t_f \mu_f}{S} \qquad (3.36)$$

where t_f and μ_f are the thickness and permeability constant of the full-scale plate, and S is the scale factor of the model. Since most of the ship's magnetic material is composed of thin plates, permeability-thickness modeling accounts for a large portion of the model's design. In the past, ship plating has typically been specified by the weight per unit area of the material (for example, pounds per foot^2) instead of its thickness. This parameter can be converted to thickness and used in the permeability-thickness scaling equation to give

$$t_m \mu_m = \frac{W_a \mu_f}{d\,S} \qquad (3.37)$$

where W_a is the weight per unit area of the plate material, and d is its density

For objects that have two dimensions much smaller than the third, their permeability $\times$ cross-sectional area is scaled. Permeability-area modeling is applied to a ship's beams and girders. For example, a ship's I-beam is modeled as a flat bar with a maximum cross-sectional dimension equal to the height of the I-beam's web. The thickness of the flat bar used to approximate the magnetic characteristics of the I-beam can be computed from

$$t_f = \frac{W_l}{h\,d} \qquad\qquad (3.38)$$

where h is the height of the I-beam's web, W_l is the linear density of the beam (such as pounds per linear foot), and the other terms have been defined earlier. Placing (3.38) into (3.36) yields

$$t_m \mu_m = \frac{W_l \mu_f}{h\,d\,S}. \qquad\qquad (3.39)$$

While the thickness of the model bar used to represent the full-scale I-beam is given by (3.39), its width (or height) should be set equal to h/S.

Shipboard items of concentrated magnetic mass whose three dimensions are of the same order of magnitude produce signatures that are dependent more on their shape rather than on permeability. All items exterior to the vessel's hull should be accurately reproduced on the model, although smaller internal machinery and equipment items are of secondary importance since they are shielded by the hull and rarely have appreciable extent in any one direction. The detail placed in modeling shielded structures has little effect on the magnetic characteristics of the overall model, allowing internal equipment and machinery items to be reproduced as boxes approximating the size and shape of the full-scale item. However, the magnetic mass of an internal item can influence the loop-effects of the ship's degaussing coils. Therefore, enough partitions are built into the item boxes to reproduce the scaled-down weight of the actual structure.

The ampere-turns of a ship's degaussing coils are scaled directly on the physical model. If $N_f I_f$ represents the full-scale ampere-turns of a vessel's degaussing loop, then the number of integer turns on the scale model N_m is computed from

$$N_m = \frac{N_f I_f}{S\,I_m} \qquad\qquad (3.40)$$

where I_m is the maximum current allowed for the model's degaussing loop set either by the power supply capacity or the conductor size. Typically, a scale model's degaussing coils are wound from #18 or #20 AWG stranded-copper wires with a Teflon jacket. The leads from each coil are twisted two to three times per inch, and are labeled and connected to a terminal strip mounted on the outside of the model.

By using (3.36)–(3.39) to scale the magnetic properties of a naval vessel, the model can be constructed with materials of workable thicknesses by applying layers of thin low-carbon steel plated with tin. Magnetic ship models are mainly constructed of electro-tin-plate with a bright reglow finish because of its good soldering and rust-inhibiting qualities. Tin-plate is procured in bundled sheets ranging in thickness from 0.15 up to 1.27 mm. Upon receipt of the material, the permeability-thickness of each sheet is individually measured and separated into groups with variances of no more than ±2.5%, and are assigned a single-, double-, or triple-letter label designating its ply. The components of the scale model are built up of one or more layers of tin-plate until the desired scaled permeability-thickness is reached.

Although nonmagnetic portions of the ship need not be included in the model, many man-hours of work by artisans are required to construct a magnetic physical scale model of a typical steel hull naval vessel. Some of the ship's structure and components that are reproduced on the model include not only the hull, superstructure, platforms, decking, bulkheads, frames, weapons, and internal machinery items; but it must also contain detailed items such as propeller shafts and shaft tubes, struts, rudders, bilge keel, anchor and chain, and protective structure for the sonar to name a few. Intermixed with these model details is the installation of the vessel's degaussing coils and their leads. The need for such detail in magnetic scale models has been demonstrated and validated over a period of 60 years.

Although it is desirable to make the magnetic ship model's scale as large as possible, the maximum weight and size of the model is set by the physical limits of the laboratory testing facility. Naval magnetic testing facilities can typically maneuver models up to about 400 kg in weight and about 3.5 m in length. As a result, model scales used for modern surface ships and submarines range from approximately 1:30 to 1:150 depending on the class.

A magnetic physical scale model testing facility must be able to generate a field to simulate any geomagnetic location on the earth's surface. To do so requires that the laboratory's field generating coils be capable of canceling the local earth's magnetic field in all directions, while having enough ampere-turn capacity to reproduce the terrestrial field at any magnetic latitude (including the southern hemisphere), and for any magnetic heading in that location. As expected, high uniformity and stability of the facility's generated field is a requirement. Typically, nonuniformity of a laboratory's coil system runs less than 0.05–0.5% of the value on its primary axis when measured in any direction. Uniformity of the applied field must be maintained over the entire volume of the active testing region located inside the building's coil system. Model laboratory facilities have been constructed to produce rectangular uniform testing volumes with dimensions as large as 3 m × 3.5 m × 12 m. To achieve high field uniformity over a large volume requires that the facility's field generating cables be designed with the proper distribution of segmented turns, and rigidly installed with high precision.

Some prevision must be made to compensate for geomagnetic fluxuations inside the laboratory. This can be accomplished by using an *earth's field reference* sensor to subtract variations in the background field from measurements taken on the scale model. The earth's field reference must be located far enough away from the facility so as not to be influenced by the model's magnetic signature or the field generated by the building's coil system; otherwise, this latter field must be correlated against the coil current and subtracted. Ideally, the earth's field reference sensor should be housed in a temperature-stabilized enclosure to prevent instrument drift from contaminating model signature measurements taken over a long span of time.

To obtain the most accurate measurements of a scale model's signature, a line of triaxial fluxgate magnetometers should be mounted on a ridged beam underneath the model or in a circle surrounding it. The model is hung from a mobile support trolley that moves it continuously in a straight line past the sensor array as they record measurements of its field signatures. Location of the model relative to the sensors is also accurately measured and recorded. This procedure results in a large number of discrete field points sampled over a cylindrical or planar surface that extends ideally a *ship's-length* before the bow and after the stern. It is desirable to start and stop the model far enough from the sensor array for its signature to be near the baseline level of the background-inducing field.

In addition to the laboratory's uniform field coil system, it should also be fitted with a small solenoid that can magnetize and demagnetize the scale model. For demagnetizing it, a slowly damped oscillating magnetic field is applied to the model while holding a near-zero background field. To magnetize the model, a bias field can be turned on in the desired direction while applying the large slowly damped oscillating field. A scale model facility's magnetizing/demagnetizing solenoid should be designed with the capacity to apply at least 2 mT of bias field, simultaneously with an oscillating magnetizing field of at least 7 mT.

Care must be taken in selecting the site and materials for constructing a magnetic testing facility. The building's location must be selected so as to maximize the distance to structures that contain steel, power machinery, and electrical distribution stations, and away from vehicular traffic such as roads and highways that can introduce periodic interference. Also, it is necessary to take a magnetic survey over the proposed site for the new facility, and should include a large buffer zone around it. The purpose of the survey is to determine if there is any natural or man-made magnetic material buried in the area that might be producing magnetic field gradients larger than the facility's uniformity requirement. If this is the case, either the material or the proposed site for the facility should be moved.

After selecting a suitable nonmagnetic location for the facility, its structure and support equipment must be specified and constructed from nonmagnetic material (wood and plastic). The use of nonmagnetic but electrically conducting metals, such as aluminum and copper, should be minimized in case the laboratory might be used for AC field measurements. It is easy for physically small but magnetically significant components to slip into the facility during construction as part of the heating and air conditioning, water pipes and bathroom fixtures, electrical conduit and breaker boxes, etc. Special care must be taken to specify the use of nonmagnetic gravel and cement in any concrete poured in or near the building. As a precaution, a sample from each batch of concrete should be screened and the entire load rejected if it does not meet the magnetic specification. It can be very expensive to correct a mistake of this type.

Before experimenting with any scale model, the entire testing system and sensor configuration should be calibrated with a magnetic source. A small magnetic dipole loop or short solenoid is mounted on the overhead trolley where the scale ship model would normally be attached. The electrical leads powering the source are twisted and supported so as not to interfere with the experiment. The source is energized with a stable DC current, which is monitored while its signatures are being measured. As the trolley moves the calibrated source in a straight line past the sensor array, its relative location and magnetic fields are continuously sampled. Comparing the dipole's signature as measured by each sensor axis against those computed theoretically will indicate those instruments that may have been misspositioned or have inaccurate readings, and will verify the accuracy of the overall experimental setup.

After calibrating and verifying the experimental setup, the dipole is replaced with the magnetic scale ship model. Care must be taken in leveling the model and in measuring its height above the sensor array's reference point. The model is then demagnetized (*depermed*), and the building's triaxial inducing fields set to that specified in the test plan. One complete pass of the model past the sensor array, over the full length of the trolley rail, while measuring its field is called a *ranging*. Repeatability of the experiment is checked at this point by reranging the model under the same test conditions. If called for by the test plan, the model can be magnetized along any direction using the facility's bias coils and high-power magnetizing/demagnetizing solenoid. The experimental data should be analyzed shortly after it is collected to ensure its quality and to avoid costly retesting.

REFERENCES

[1] J. D. Jackson, *Classical Electrodynamics*, 3rd ed. Hoboken, NJ: Wiley, 1999, pp. 201–203.

[2] L. Frumkis and B. Z. Kaplan, "Spherical and spheroidal shells as models in magnetic detection," *IEEE Trans. Magn.*, vol. 35, no. 5, pp. 4151–4158, Sep. 1999.

[3] F. E. Baker and S. H. Brown, "Magnetic induction of spherical and prolate spheroidal bodies with infinitesimally thin current bands having a common axis of symmetry and in a uniform inducing field; a summary," David W. Taylor Naval Ship Res. Dev. Center, West Bethesda, MD, Tech. Rep. DTNSRDC-81/014, Jan. 1982.

[4] C. H. Sinex, "Dipole and quadrupole analysis of magnetic fields," Johns Hopkins Univ. Appl. Phys. Lab., Laurel, MD, Tech. Rep. POR-3038, Dec. 1971.

[5] W. R. Smythe, *Static and Dynamic Electricity*. New York: McGraw-Hill, 1968, pp. 290–291.

[6] S. H. Brown and F. E. Baker, "Magnetic induction of ferromagnetic spherical bodies and current bands," *J. Appl. Phys.*, vol. 53, pp. 3981–3990, 1982.

[7] F. E. Baker and S. H. Brown, "Magnetic induction of ferromagnetic prolate spheroidal bodies and infinitesimally thin current bands," *J. Appl. Phys.*, vol. 53, pp. 3991–3996, 1982.

[8] X. Brunotte, G. Meunier, and J. P. Bongiraud, "Ship magnetizations modeling by the finite element method," *IEEE Trans. Magn.*, vol. 29, no. 2, pp. 1970–1975, Mar. 1993.

[9] F. Le Dorze, J. P. Bongiraud, J. L. Coulomb, and P. Labie, "Modeling of degaussing coils effects in ships by the method of reduced scalar potential jump," *IEEE Trans. Magn.*, vol. 34, no. 5, pp. 2477–2480, Sep. 1998.

[10] B. Klimpke (2006, Mar.). A hybrid magnetic field solver: Using a combined finite element/boundary element field approach. *Sensors* [Online]. Available: http://www.sensorsmag.com/articles/0504/14/main.shtml

[11] "A short history of degaussing," Bureau Ordnance, Washington, DC, Tech. Rep. NAVORD OD 8498, Feb. 1952.

[12] S. Fry and C. E. Barthel, Jr., "Design and construction of the magnetic model of the DE-52," Naval Ordnance Lab., Washington, DC, Tech. Rep. NOLR 811, Jan. 1947.

CHAPTER 4

Semiempirical Models

Unlike the first-principal mathematical and physical scale models discussed in the previous chapter, semiempirical models are used to extrapolate magnetic signatures of surface ships and submarines from their measurement-environment and geometry to a threat-environment. The signatures to be extrapolated can be either from at-sea measurements on a full-scale vessel, or they can come from a scale model tested in the laboratory. In either case, a mathematical model of the ship is used to perform the extrapolation.

There are two steps in the semiempirical model extrapolation of magnetic signatures. First, an equivalent source mathematical model of the ship is formulated that can be used to compute triaxial magnetic signatures along any line or surface outside the vessel's hull. This is called a forward-model. The source strengths of the forward-model are typically computed with an inverse-model that uses as input actual field measurements and the ship-to-sensor geometry. In this chapter, several types of forward-models will be described that can also be used in inverse-model computations. In addition, the inherent instabilities of inverse-models will be discussed on a theoretical basis, along with techniques that can be used to regularize the process.

4.1 FORWARD MODELS

The simplest mathematical model that can be used to describe the magnetic signature of a ship or submarine is a spherical dipole, also called a point dipole. The equations relating triaxial magnetic point dipole moments to their triaxial field components B_x, B_y, B_z are given in the *SI* system by

$$B_x = \frac{\mu_0[m_x(2x^2 - y^2 - z^2) + 3m_y x\, y + 3m_z x\, z]}{4\pi(x^2 + y^2 + z^2)^{\frac{5}{2}}} \tag{4.1}$$

$$B_y = \frac{\mu_0[3m_x x\, y + m_y(2y^2 - x^2 - z^2) + 3m_z y\, z]}{4\pi(x^2 + y^2 + z^2)^{\frac{5}{2}}} \tag{4.2}$$

$$B_z = \frac{\mu_0[3m_x x\, z + 3m_y y\, z + m_z(2z^2 - x^2 - y^2)]}{4\pi(x^2 + y^2 + z^2)^{\frac{5}{2}}} \tag{4.3}$$

where (m_x, m_y, m_z) are the longitudinal, athwartship, and vertical spherical dipole moments of the ship, respectively, and all other parameters have been defined previously. The signatures of all magnetic bodies can be represented by (4.1)–(4.3) if the field measurements are taken far enough away. The range at which a ship's signature can be accurately reproduced by the point dipole equations alone is called the *far-field* of the vessel. (This use of the term far-field should not be confused with that ascribed to the radiation region of a time-harmonic dipole.) Conversely, a ship or submarine's magnetic *near-field* encompasses the region closer to their hulls where the signature is more complex, and requires additional source terms to accurately describe it.

As the length of a surface ship or submarine is generally several times larger than its beam, prolate spheroidal sources are typically used for modeling their magnetic signatures. A prolate spheroidal dipole model will more accurately reproduce a naval vessel's signatures at distances much closer than a spherical dipole. As discussed in the previous chapter, a spherical dipole moment m can be equated to a prolate spheroidal moment M using $M = 3m/c^2$. The magnetic field components of a triaxial prolate spheroidal dipole source, whose long direction is aligned with the z-axis, can be expressed as

$$B_x = \frac{3\mu_0}{4\pi c^3} \left\{ m_x \left[\frac{1}{4} \ln \left(\frac{\xi+1}{\xi-1} \right) - \frac{\xi}{2\left(\xi^2-1\right)} + \frac{x^2\xi}{r_1 r_2 \left(\xi^2-1\right)^2} \right] + m_y \left[\frac{xy\xi}{r_1 r_2 \left(\xi^2-1\right)^2} \right] \right.$$
$$\left. + m_z \left[\frac{cx\eta}{r_1 r_2 \left(\xi^2-1\right)} \right] \right\} \tag{4.4}$$

$$B_y = \frac{3\mu_0}{4\pi c^3} \left\{ m_x \left[\frac{xy\xi}{r_1 r_2 \left(\xi^2-1\right)^2} \right] + m_y \left[\frac{1}{4} \ln \left(\frac{\xi+1}{\xi-1} \right) - \frac{\xi}{2\left(\xi^2-1\right)} + \frac{y^2\xi}{r_1 r_2 \left(\xi^2-1\right)^2} \right] \right.$$
$$\left. + m_z \left[\frac{cy\eta}{r_1 r_2 \left(\xi^2-1\right)} \right] \right\} \tag{4.5}$$

$$B_z = \frac{3\mu_0}{4\pi c^3} \left\{ m_x \left[\frac{cx\eta}{r_1 r_2 \left(\xi^2-1\right)} \right] + m_y \left[\frac{cy\eta}{r_1 r_2 \left(\xi^2-1\right)} \right] + m_z \left[\frac{-1}{2} \ln \left(\frac{\xi+1}{\xi-1} \right) + \frac{c^2\xi}{r_1 r_2} \right] \right\} \tag{4.6}$$

where m_x, m_y, and m_z are equivalent spherical dipole moments, and ξ, η, c, r_1, and r_2 have all been defined in the Appendix in terms of x, y, and z. It must be pointed out that these equations have been formulated in terms of equivalent spherical dipole moments so that in the far-field (4.1)–(4.3) generate the same signatures as (4.4)–(4.6), respectively.

To accurately reproduce a ship's magnetic signature at distances very close to its hull requires the inclusion of higher order terms in the spherical or prolate spheroidal multipole expansions. The equations for the complete series expansion of spherical and prolate spheroidal harmonics (dipole, quadrupole, octupole, etc.) and their application to ship signature modeling can be found in [1], along with numerical examples showing faster convergence of the spheroidal series over the spherical for this application. Faster convergence is important when using these formulations for inverse-modeling. The less harmonic terms required for accurate reproduction of a near-field signature, the fewer measurements and sensors are needed to solve for all the source strengths in the expanded series.

Although spherical and spheroidal harmonic models reproduce magnetic signatures using orthogonal bases functions, a mathematically desirable condition, hybrid models have been used extensively in ship signature analysis. One of the earliest mathematical magnetic ship models was comprised of a single triaxial spheroidal dipole, two spheroidal quadrupoles, plus four to six triaxial spherical dipoles placed along the vessel's centerline. It was found that the spheroidal sources accounted for the bulk of the hull's contribution to the ship's signature, while the point dipoles reproduced the effects of internal structure and nonuniformities in the hull. The advantage of this model was its ease in implementation.

There are several other types of mathematical magnetic ship models that have been considered in the past. One such model uses only triaxial spheroidal dipoles positioned at known locations of concentrated magnetic mass. Another model represents the magnetic fields of the vessel with a continuous line of magnetization described by polynomials. In one case, a line of several hundred triaxial point dipoles was placed along the centerline of the ship.

As an example in how to implement a forward-model, consider the case where the magnetization of a ship or submarine can be represented by 11 vertical spherical magnetic dipoles located along its centerline. The dipole distribution used in this example is shown at the top of Fig. 4.1. Their maximum source strengths in this case have been normalized to 1, and they are distributed uniformly along the vessel between +1 and −1, which represent normalized half-lengths of the ship. The vertical magnetic field signature was computed under the keel at a scaled depth of 1 as shown at the bottom of the figure. The location of the dipoles relative to the signature is as indicated.

To compute the signature shown at the bottom of Fig. 4.1 requires the creation of two matrices. The first array should contain the 11 vertical dipole moments $[m_z]$, and the second their longitudinal location $[x']$. The entries in the column matrix $[m_z]$ run from 0 up to 1 and back down to 0 as shown in the figure, while those in the source location column matrix

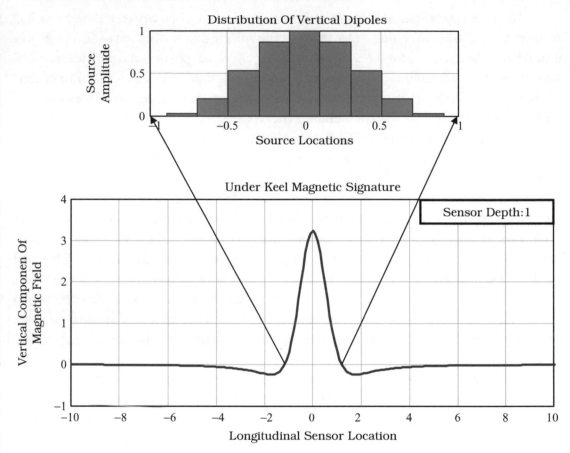

FIGURE 4.1: Forward-model prediction of the vertical magnetic field signature from a linear distribution of vertical point dipoles.

$[x']$ range from -1 to $+1$ in ten equal steps. The field at each x_i coordinate, which, in this example, runs from -10 to 10 in equal steps, can now be computed from (4.3) by setting $m_x = m_y = y = 0$ and $z = 1$ to give

$$B_i = \sum_{j=1}^{11} c_{i,j} m_j \qquad (4.7)$$

where

$$c_{i,j} = \frac{\mu_0}{4\pi} \frac{2 - \left(x_i - x'_j\right)^2}{\left(\left(x_i - x'_i\right)^2 + 1\right)^{\frac{5}{2}}}$$

In matrix form, (4.7) can be represented as

$$
\begin{bmatrix}
c_{1,1} & c_{1,2} & \cdots & c_{1,10} & c_{1,11} \\
c_{2,1} & c_{2,2} & \cdots & c_{2,10} & c_{2,11} \\
\vdots & \vdots & \ddots & \vdots & \vdots \\
c_{i-1,1} & c_{i-1,2} & \cdots & c_{i-1,10} & c_{i-1,11} \\
c_{i,1} & c_{i,2} & \cdots & c_{i,10} & c_{i,11}
\end{bmatrix}
\begin{bmatrix}
m_1 \\
\vdots \\
m_{11}
\end{bmatrix}
=
\begin{bmatrix}
B_1 \\
B_2 \\
\vdots \\
B_{i-1} \\
B_i
\end{bmatrix}
\tag{4.8}
$$

or more succinctly,

$$
[c]\ [m_z] = [B_z]
\tag{4.9}
$$

The magnetic component B_z represents data that might be collected on a full-scale vessel or from measurements taken in a laboratory on a scale magnetic model.

In the example considered earlier, the discrete nature of the 11 point dipoles is not apparent in the overall signature that they produced. The field contribution from each point source is as if its magnetization was smoothed out and blended together with the others. This characteristic is reminiscent of filtering.

The blurring of individual source contributions to the total signature of a ship can be explained by examining the integral solution to the Poisson equation. The Poisson integral can be formulated as a convolution of the Green's function, $g(x - x')$, with the vessel's magnetization distribution, $m(x')$, to produce the magnetic field signature, $h(x)$, as given for the one-dimensional case by

$$
h(x) = \int_a^b g(x - x')\, m(x')\, dx'
\tag{4.10}
$$

where a and b are the coordinates of the bow and stern. Using the convolution theorem, (4.10) can be written in the frequency domain as

$$
H(\omega) = G(\omega)\, M(\omega)
\tag{4.11}
$$

where $H(\omega)$, $G(\omega)$, and $M(\omega)$ are the Fourier transforms of $h(x)$, $g(x)$, and $m(x)$, respectively, and ω represents spatial frequency instead of temporal.

As suggested by (4.11), the Green's function acts as a filter for a ship or submarine's magnetization distribution, producing magnetic field signatures with lower spatial frequency content. A qualitative example is shown in Fig. 4.2. The spatial frequency content of a magnetization distribution is given on the left, which when multiplied by the Green's function spectrum in the center, produces the spectrum of the magnetic field signature on the right. Comparison of

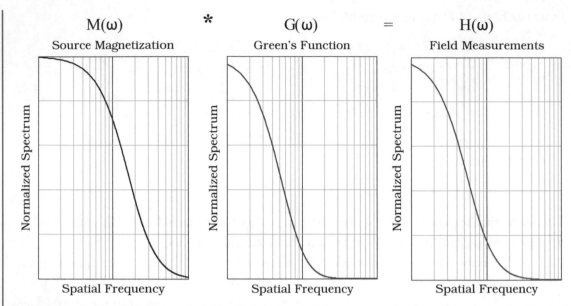

FIGURE 4.2: Graphical example of predicting the spatial spectrum of a magnetic signature by filtering the spectrum of the ship's magnetization distribution with that of the Green's function.

the higher frequency energy in the magnetization spectrum to that of the magnetic field shows that it has been attenuated by the low-pass characteristics of the Green's function. As the field measurement distance approaches the source (in this case, the ship), the spectral shape of the Green's function changes (see Fig. 4.3), and allows higher spatial frequencies of the source's magnetization to pass into the magnetic signature. This characteristic is due to the underlying physics of the problem, and is independent of the mathematical model selected to represent the vessel.

In general, the best equivalent source model is a matter of personal preference. If a model reproduces the magnetic signature of a ship with the desired accuracy then it is adequate. Although some models may be more or less unstable when used in the inverse mode, the physics of the inverse problem subjects all of them to the same uniqueness issues to be discussed next.

4.2   INVERSE MODELS

The purpose of an inverse-model is to determine the equivalent source strengths needed as input to forward-models. Once the source strengths have been evaluated, the forward-model can be used to predict the vessel's magnetic field in other environments and ship-sensor encounter geometries, including those of specific threat scenarios. Therefore, the forward and inverse models are tightly linked and, in fact, are the same, differing only in how they are used.

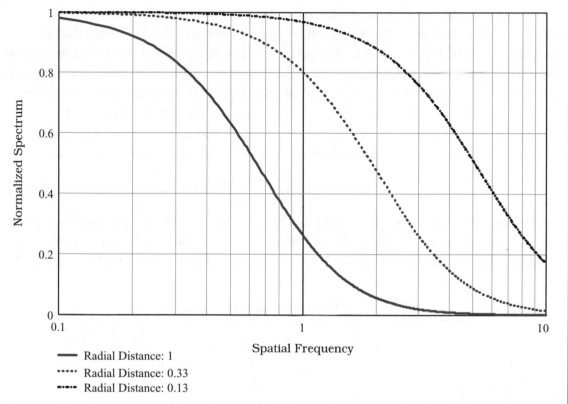

Radial Distance: 1
Radial Distance: 0.33
Radial Distance: 0.13

FIGURE 4.3: Variation of the Green's function spatial filter with ship-to-sensor offset.

The inputs to an inverse-model are the ship or submarine's magnetic field measured by one or more sensors, along with the three-dimensional coordinates and angular direction of the ship's axis relative to the sensors'. Field measurements can be taken with the vessel moored over an array of magnetic instruments, or more typically, the ship sails past a line of sensors called a *range* that sample its signature as a function of time. In the latter case, tracking systems are used to convert the measured time-domain signatures into the spatial domain. This *ranging* process produces a complete two-dimensional spatial array of triaxial field measurements from a single line of sensors installed transverse to the ship track.

Once the array of magnetic field data has been recorded, along with the three-dimensional coordinates of each measurement sensor relative to the ship, an inverse-model is used to compute the equivalent source strengths that will reproduce the signatures. Generally, the constrained linear least-squares minimization algorithm is used to compute the source strengths from the measured data. The forward-model example given earlier will be used to explain the inversion process, and the instabilities associated with it.

In this inversion example, the signature at the bottom of Fig. 4.1 will now represent that of a measured full-scale vessel or a scale model. The objective here is to determine the 11 source strengths $[m_z]$, which, in this case, are unknowns. The inputs to the inverse-model are the column array of vertical magnetic field values $[B_z]$ measured at sensor locations given by $y = 0$, $z = 1$, and the column array of longitudinal coordinates $[x]$. The equation to be solved is once again (4.9), but, in this case, the terms in the column matrix $[m_z]$ are the unknowns.

The least-squares minimization algorithm will be used to solve the overdetermined set of equations given by (4.9). The standard least-squares solution can be computed from

$$[m_z] = \left[[c]^T [c] \right]^{-1} [c]^T [B_z]. \tag{4.12}$$

If the noiseless signature at the bottom of Fig. 4.1 is used in (4.12), the correct (original) distribution of vertical dipoles at the top of the figure is recomputed. However, if a small random noise of the order of $\pm 1\%$ of $[B_z]$ (see the top of Fig. 4.4) is added to the array of

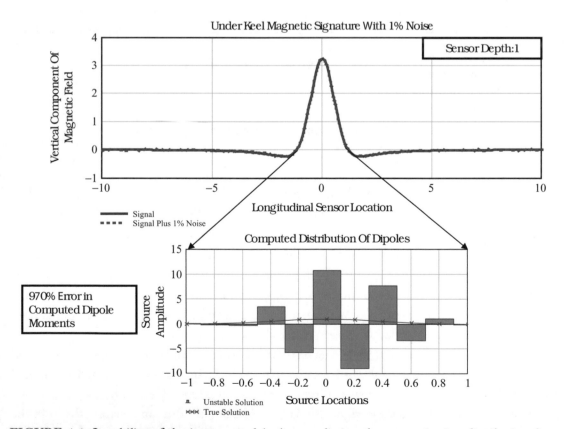

FIGURE 4.4: Instability of the inverse-model when predicting the magnetization distribution from signature data with a 1% additive noise.

measurements, then the dipole distribution computed with (4.12) is in error by as much as 970% as shown at the bottom of Fig. 4.4. It is clear that the inversion-model can be extremely unstable, and requires further investigation. A hint to the cause of the instability problem can be gleaned from the convolution equation (4.11). Rewriting (4.11) to solve for $M(\omega)$ gives

$$M(\omega) = H(\omega) / G(\omega). \qquad (4.13)$$

If a small broadband noise $\Xi(\omega)$ is added to the field measurements $H(\omega)$, then (4.13) becomes

$$M(\omega) = H(\omega) / G(\omega) + \Xi(\omega) / G(\omega). \qquad (4.14)$$

As shown in Figs. 4.2 and 4.3, $G(\omega)$ approaches zero at higher frequencies. Therefore, any small measurement noise, $\Xi(\omega)$, or experimental error that is present at frequencies where $G(\omega)$ is small will result in significantly amplified noise. This situation is shown qualitatively in Fig. 4.5. The distribution of magnetization computed with the inverse model results in the correct solution plus a very large error term with high spatial frequency content. These characteristics are clear in the unstable solution of the example shown in Fig. 4.4.

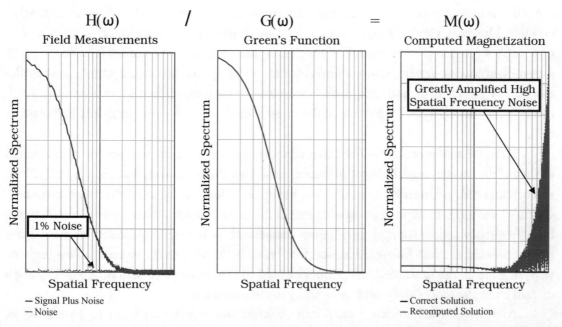

FIGURE 4.5: Graphical example of an inverse-model's instability as manifested in its spatial frequency domain.

Due to the importance of the inverse-model instability, a more mathematically rigorous explanation is warranted. Beginning with an identity proved in [2] and given by

$$\lim_{\omega \to \infty} \int_a^b g(x - x') A \sin(\omega x') \, dx' = 0 \qquad (4.15)$$

where A can be arbitrarily large, (4.10) can be written as

$$h(x) + \varepsilon(x) = \int_a^b g(x - x') \left[m(x') + A \sin(\omega x') \right] dx' \qquad (4.16)$$

where $\varepsilon(x)$ can be made small, even for large A, provided ω is also large. If $h(x) + \varepsilon(x)$ represents a measured magnetic signature with additive noise, then solving the inverse problem for the ship's magnetization will yield the true solution plus a large amplitude, high frequency, error term given by $A \sin(\omega x')$. There are many combinations of A and ω in (4.16) that will produce noise on the order of $\varepsilon(x)$. Therefore, the conclusion is that in the presence of noise or experimental error, the solutions to inverse-models are nonunique regardless of their specific formulation.

The arguments given earlier suggest that to avoid the nonuniqueness instability of inverse-models, the magnetic signatures to be used as its input should be measured as close to the vessel as possible. However, very high-order source terms are required to obtain an accurate reproduction of a ship's signature close to its hull. By including many degrees of freedom (source terms) in a near-field inverse-model, sensor numbers, ship tracking accuracy requirements, and model complexity increases. In addition, models with an excessive number of source terms may still result in an unstable solution in spite of the measurement sensors' close proximity to the ship's hull.

If the forward model is confined to predicting signatures at distances greater than or equal to that of the inverse-model's input measurements, then the extrapolated signatures will still be smooth and stable. Under these conditions, the upper cutoff frequency of the forward-model's Green's function is lower than or equal to that of the inverse-model. Therefore, all the incorrect high-frequency source strength oscillations introduced by the inverse-model are filtered out by the forward-model's Green's function filter with its lower cutoff frequency. This explains why, in general, the forward-model cannot be used to extrapolate a vessel's magnetic signature inward from the sensor array used as input by the inverse-model.

At times, an indication of the relative distribution of magnetization in a naval vessel is needed to isolate problem areas during the design and calibration of signature reduction systems. In this case, the stability of a ship's inverse-model is an issue that must be addressed.

Since all inverse-models have inherently nonunique solutions, additional information must be introduced into the problem to select out of all the possible solutions the one that is closest to that desired for the application at hand.

There are many techniques that have been developed for stabilizing the solutions to problems of *indirect measurement*. Some of the more classic methods used in radio astronomy and radiometry, seismology, optics, geophysics, acoustics, and atmospheric sensing can be found in [3]. Criteria used to select the desired solution out of the nonunique set include the smoothest solution condition (minimum first- or second-order derivative), the solution closest to a desired distribution, and the minimum energy solution. Typically, inverse modeling of surface ship and submarine signatures uses the minimum energy criteria.

The minimum energy criteria, as applied to magnetic field inverse-models, requires that the sum of the square of the computed source strengths be a minimum, while simultaneously producing a good reproduction of the vessel's signature. The mathematical basis of the minimum energy constraint is derived in [3], and its implementation is straightforward. The minimum

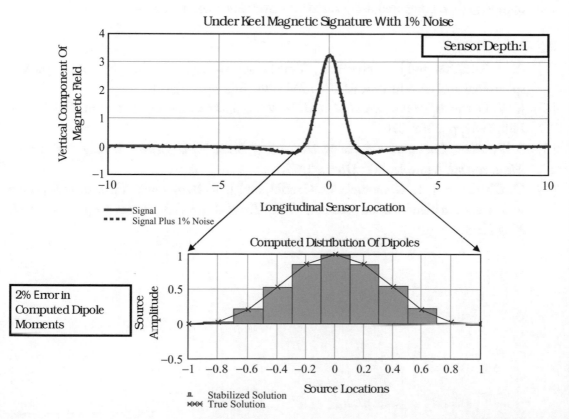

FIGURE 4.6: Stabilized inverse-model computed from signature data with a 1% additive noise.

energy constraint is inserted into signature inversion problem by modifying (4.12) to give

$$[m_z] = \left[[c]^T [c] + [I]\, \alpha\right]^{-1} [c]^T [B_z] \qquad (4.17)$$

where $[I]$ is the identity matrix, and α is a weighting factor. In practice, the α term (sometimes called a *damping factor*) is adjusted empirically to produce the minimum energy solution that also reproduces the input signature to within error bounds on the order of the measurement system's noise level. If (4.17) instead of (4.12) is now used to invert the noisy signature at the top of Fig. 4.5, the smooth solution at the bottom of Fig. 4.6 is computed. This stabilized solution is close to our desired source distribution (2% error), and reproduces the original input signature within its noise level as shown at the top of Fig. 4.6.

Recently, a method has been reported where the hull material's characteristics are used to introduce the additional information needed to solve the inverse problem [4]. This technique does not use the numerical stabilization approaches described earlier. Instead, it reposes the problem by taking field measurements close to the ship's hull and linking the many near-field source terms according to the physics of magnetization, reducing the number of unknowns. The solution is then computed using standard matrix inversion methods.

REFERENCES

[1]   A. V. Kildishev and J. A. Nyenhuis, "Zonal magnetic signatures in spherical and prolate spheroidal analysis," in *Proc. MARELEC 1999*, Jul., pp. 231–242.

[2]   J. W. Dettman, *Mathematical Methods in Physics and Engineering*. New York: McGraw-Hill, 1962, pp. 369–370.

[3]   S. Twomey, *Introduction to the Mathematics of Inversion in Remote Sensing and Indirect Measurement*. Mineola, NY: Dover, 1977.

[4]   O. Chadebec, J. L. Couloumb, G. Cauffet, and J. P. Bongiraud, "How to well pose a magnetization identification problem," *IEEE Trans. Magn.*, vol. 39, no. 3, pp. 1634–1637, May 2003.

CHAPTER 5

Summary

Mathematical and physical scale models have been used successfully for over 60 years to predict and extrapolate the ferromagnetic signatures of surface ships and submarines. The two general classes of ship models are the forward and inverse. The forward class can be broken down further into physical scale and math models, while inverse models are semiempirical in nature.

First-principal models fall under the forward class and use the constituent parameters of a ship to predict its magnetic signatures. The input parameters to first-principal models are the geometry of the hull, internal structure, and machinery and equipment items; along with their magnetic properties. First-principal models are used primarily to predict the uncompensated and compensated fields of ships that are still under design. Adjustments in the design of a vessel's signature reduction system are made using first-principal models to optimize its performance, and to minimize its cost and impact on the ship and its systems.

Magnetic physical scale models of surface ships and submarines are the older of the first-principal modeling techniques, originating during World War II. Two techniques have been used to design physical scale magnetic models of naval vessels. Thickness-scaling is a straightforward scaling of all ship dimensions, including the thickness of its hull, while using the same steel in the construction of the scale model as to be used for the full-scale vessel. For some ship classes, the scaled hull thickness is too small to be machined to precise specifications without deforming during construction and subsequent testing. Under these conditions, permeability-thickness modeling is employed where the scaled permeability-thickness product of the model's hull is maintained even though it may be constructed from materials with magnetic and mechanical properties different than those of the full-scale ship. Magnetic physical scale models of surface ships and submarines are still used today for designing advanced degaussing systems.

Mathematical first-principal models range from simple analytic formulations of the in-duced magnetization and degaussing loop-effects for idealized hull shapes, on up to detailed numerical simulations of the magnetic characteristics of complex hull shapes, internal struc-tures, and machinery items. The mathematics of generalized coordinate systems and their vector operators have been presented, along with example applications to the spherical and prolate spheroidal system. In addition, the advantages and drawbacks of applying the finite element technique to numerically modeling the magnetic signatures of ships were given, along with suggestions for optimizing its implementation.

Semiempirical models are used to extrapolate magnetic signatures of surface ships and submarines from their measurement-environment and geometry to threat-environments. This technique involves a two-step process. First, an equivalent source mathematical forward-model of a ship is constructed that can compute triaxial magnetic fields at any location around the hull. The strengths of the equivalent sources are evaluated with an inverse-model that uses as input actual field measurements and the ship-to-sensor geometry. Although the source strengths computed with the inverse-model are inherently unstable and require regularization, correct and stable signatures can always be regenerated at distances greater than or equal to the array of input field sensors. If these regenerated signatures are compared to actual field measurements taken with the same geometrical sensor configuration, the accuracy of the semiempirical model can be quantified.

Verification and validation of mathematical and physical scale models are an absolute requirement before using them. The purpose of model verification is to check that Maxwell's equations have been applied to the problem correctly, to check for errors in the mathematical formulation of solutions to these equations, and to verify the accuracy of subsequent manual computations or computer programs that numerically calculate the solutions. Verification can be in the form of an independent peer review of a model's formulation and implementation, comparisons to simplified and accepted solutions of well-known classic problems, and by comparing signatures predicted from different models developed by other investigators.

One level of model verification was demonstrated in Chapter 3. In the first example, the induced signatures predicted by an analytic formulation of a spherical shell were compared to that of a prolate spheroidal shell with its length set equal to its width. In this case, the prolate spheroidal model was checked against the solution to a classic boundary value problem. Another example of model verification is shown in Figs. 3.8 and 3.9. Here, the fields computed from the analytic formulation of an air-core degaussing loop are compared to that produced by one inside a spherical shell model of a ship's hull with its permeability constant set equal to one. Although these comparisons may seem trivial, performing this level of verification may uncover formulation errors at an early stage in the model's development where they can be easily corrected.

A higher level of model verification can be achieved by comparing magnetic signatures generated by a mathematical construct to those measured in the laboratory on a physical realization of the object being modeled. For example, the vertical signature analytically computed using (3.26) of Chapter 3 and experimentally measured [1] on a solid prolate spheroid is compared in Fig. 5.1. An 18,000-nT uniform magnetic field was applied to the longitudinal axis of the spheroid, which has a permeability constant of 80 and a major and minor axis length

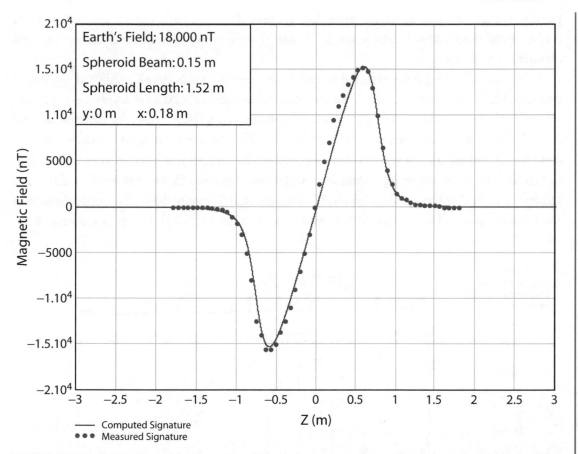

FIGURE 5.1: Comparison of the computed vertical magnetic signatures of the induced longitudinal magnetization of a solid prolate spheroid with that measured in the laboratory on a physical model.

of 1.5 m and 0.15 m, respectively. The sensor was located 0.18 m below the major axis of the spheroid. The excellent agreement between the two models shown in Fig. 5.1 does not mean that they accurately represent a ship, but it does provide confidence that the theory has been understood and applied correctly, and that proper experimental technique has been established. Verification of a model that predicts the magnetic field signatures of naval vessels must be followed by its validation against actual measurements.

The more difficult task of validation can only be achieved over an extended period of time when consistent favorable comparisons between the model and full-scale measurements have established a reliable record of accurate signature predictions. Validation is the all-encompassing process that demonstrates the model and its predictions are representations of what can be expected in the real world. In this case, magnetic fields measured on full-scale vessels compared to those predicted by simulations should fall within an accuracy band sufficiently narrow to provide confidence that signature reduction systems designed with the model will perform according to specifications. Many times the model verification is incorrectly referred to and accepted as validation.

When validating first-principal models, the comparisons between predicted signatures and actual field measurements should be part of a double-blind test. As an example, a comparison between the vertical magnetic field signatures measured on a DE 52 class destroyer escort and its scaled model's predictions is shown in Fig. 5.2 [2]. The earth's magnetic field at the test site was 52,000 nT in the vertical direction and 17,000 nT in the horizontal (north–south) direction. The lower set of signatures is a comparison between those generated by the ship's induced longitudinal magnetization (ILM) with its model's prediction. The upper signatures are a comparison of their equilibrium vertical magnetization (EVM), which is the sum of the

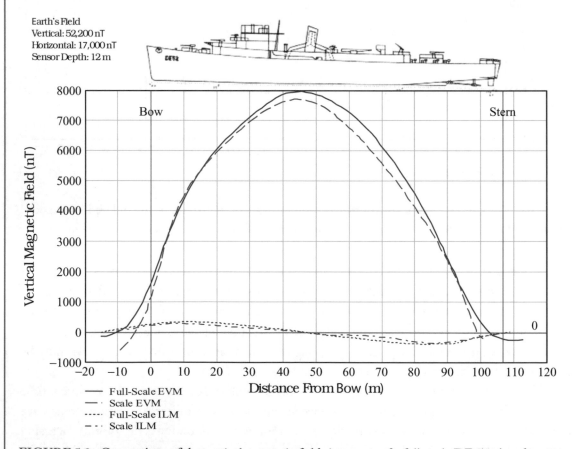

FIGURE 5.2: Comparison of the vertical magnetic field signatures of a full-scale DE 52 class destroyer escort with that of its scale model for the induced longitudinal magnetization (IVM) and the equilibrium vertical magnetization (EVM).

induced vertical magnetization (IVM) and the permanent vertical magnetization (PVM). The difference between the predicted and true signatures is less than 10% of the peak field, and is typical for scale models as validated with 60 years of favorable comparisons.

An important requirement of magnetic models is that they reliably predict the performance of signature reduction systems. Degaussing coils are the primary system used to actively cancel the magnetic fields of naval vessels. An *M-type* degaussing coil compensates the vertical magnetization of a ship. The change in a vessel's magnetic signature produced by a 1-A change in the degaussing loop's current is defined as a loop-effect. The average M-coil loop-effects at a depth of 12 m as measured on several different DE 52 class hulls are compared in Fig. 5.3 to that predicted by the scale model. The data shows that the M-coil loop-effect has about a 10% variance between ships within the class, and that the scale model prediction falls within that range.

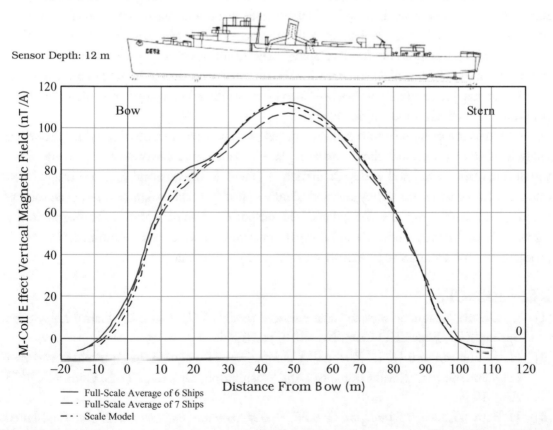

FIGURE 5.3: Comparison of the vertical component of the *M-coil* loop-effect as measured on several full-scale DE 52 class destroyer escorts with that of its scale model.

As a general rule, the signatures predicted by a properly designed and built scale model fall within the range of variation observed on different hulls within the same class. However, due to the 10% spread observed in measured signatures between different hulls that are supposed to be identical, no first-principal model can claim to be validated to better than this value.

Used properly, analytic, numerical, and physical scale magnetic models of surface ships and submarines are all important tools for reducing the risk and cost in designing signature reduction systems. However, the models' outputs should not be taken as the absolute truth and proof of a design. As an example, the design fault that caused the roof of Connecticut's Hartford Civic Center to collapse in 1978 was traced to flawed assumptions used as input to an elaborate computer model of its frame members [3]. Most assuredly, the application of structural theory and the programming of the complex computer software were verified as correct. However, by inputting oversimplified and incorrect parameters into the program, the model as constructed was invalid.

Modeling and simulation are not substitutes for critical thinking and engineering judgment. There are many vendors and university researchers who have developed accurate and verified finite- and boundary-element software packages. However, it is the users of these programs who are responsible for developing accurate and validated models. Placing incorrect input information into verified software does not somehow produce validated output. It is even possible to corrupt a model that has been previously validated by using it outside the range of parameters for which it was originally tested.

Models are only tools to confirm or dispel possible design errors flagged by the critical analysis of an engineer, and should never replace judgment and experience. Too many times, the outputs from numerical and scale models are taken as unequivocal fact and true without question. In the worse case, they are used blindly to justify a bad design that has been selected based on erroneous preconceived beliefs. The output of an underwater electromagnetic ship signature model, or any model, should always be questioned and seriously examined. Real-world processes are too complex to be exactly reproduced by a simulation.

REFERENCES

[1] J. A. Ford, "Magnetic signatures of ellipsoid models VII," Naval Ordnance Lab., Silver Spring, MD, Rep. EED Com. No. 4013, Jan. 1966.

[2] H. P. Hanson and G. L. Parsons, "A comparison of magnetic fields of ships and their magnetic models," Naval Ordnance Lab., Washington, DC, Rep. Tech. Com. No. 8655, Aug. 1946.

[3] H. Petroski, *Design Paradigms: Case Histories of Error and Judgment in Engineering*, 1st ed. Cambridge, UK: Cambridge University Press, 1994.

Appendix I

COORDINATE TRANSFORMATIONS AND OPERATORS

Rectangular

The rectangular or Cartesian coordinate system is the easiest and most straightforward to apply. The system for rectangular coordinates has been drawn in Fig. I.1 showing surfaces formed by holding one of the coordinates constant. The equations describing the metrics, gradient, divergence, Laplacian, and curl operators in the Cartesian coordinate system are:

Coordinates:

$$(x, y, z) \quad -\infty \leq x < \infty, \quad -\infty \leq y < \infty, \quad -\infty \leq z < \infty \qquad (I.1)$$

Metrics:

$$h_x = h_y = h_z = 1 \qquad (I.2)$$

Gradient:

$$\nabla \Phi = \hat{x}\frac{\partial \Phi}{\partial x} + \hat{y}\frac{\partial \Phi}{\partial y} + \hat{z}\frac{\partial \Phi}{\partial z} \qquad (I.3)$$

Divergence:

$$\nabla \cdot \vec{H} = \frac{\partial H_x}{\partial x} + \frac{\partial H_y}{\partial y} + \frac{\partial H_z}{\partial z} \qquad (I.4)$$

Laplacian

$$\nabla^2 \Phi = \frac{\partial^2 \Phi}{\partial x^2} + \frac{\partial^2 \Phi}{\partial y^2} + \frac{\partial^2 \Phi}{\partial z^2} \qquad (I.5)$$

Curl

$$\nabla \times \vec{H} = \hat{x}\left(\frac{\partial H_z}{\partial y} - \frac{\partial H_y}{\partial z}\right) + \hat{y}\left(\frac{\partial H_x}{\partial z} - \frac{\partial H_z}{\partial x}\right) + \hat{z}\left(\frac{\partial H_y}{\partial x} - \frac{\partial H_x}{\partial y}\right) \qquad (I.6)$$

Cylindrical

The first of the curvilinear coordinate systems is the cylindrical. The system for cylindrical coordinates has been drawn in Fig. I.2 with the typical orientation of the z-axis along the

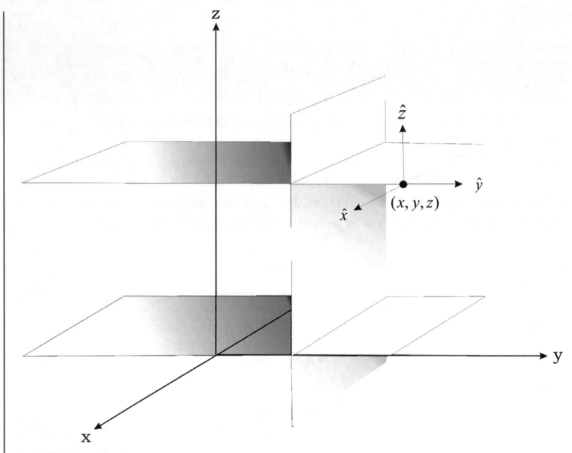

FIGURE I.1: Rectangular coordinate system.

length of the cylinder. The coordinate and vector transformations between the rectangular and cylindrical systems, and the metrics, gradient, divergence, Laplacian, and curl operators are given by

Coordinates:

$$(\rho, \varphi, z) \quad 0 \leq \rho < \infty, \quad 0 \leq \varphi \leq 2\pi, \quad -\infty \leq z < \infty \tag{I.7}$$

Metrics:

$$h_\rho = 1, \quad h_\varphi = \rho, \quad h_z = 1 \tag{I.8}$$

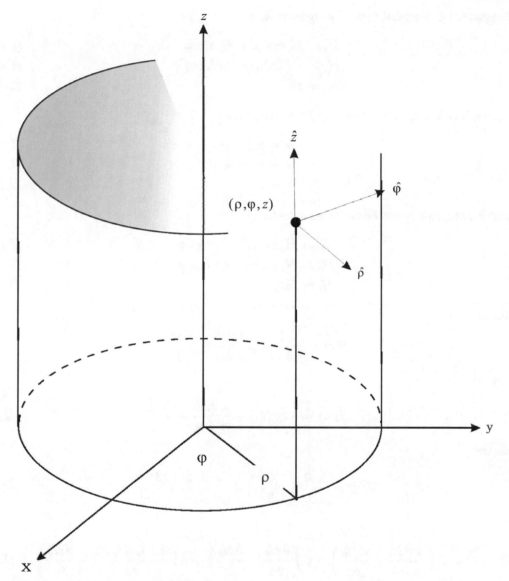

FIGURE I.2: Cylindrical coordinate system.

Rectangular to Cylindrical Coordinate Transformations:

$$\rho = \sqrt{x^2 + y^2} \qquad (I.9a)$$

$$\varphi = \tan^{-1}\left(\frac{y}{x}\right) \qquad (I.9b)$$

$$z = z \qquad (I.9c)$$

Rectangular to Cylindrical Vector Transformations:

$$H_\rho = H_x \cos\varphi + H_y \sin\varphi \qquad\qquad (I.10a)$$
$$H_\varphi = -H_x \sin\varphi + H_y \cos\varphi \qquad\qquad (I.10b)$$
$$H_z = H_z \qquad\qquad (I.10c)$$

Cylindrical to Rectangular Coordinate Transformations:

$$x = \rho \cos\varphi \qquad\qquad (I.11a)$$
$$y = \rho \sin\varphi \qquad\qquad (I.11b)$$
$$z = z \qquad\qquad (I.11c)$$

Cylindrical to Rectangular Vector Transformations:

$$H_x = H_\rho \cos\varphi - H_\varphi \sin\varphi \qquad\qquad (I.12a)$$
$$H_y = H_\rho \sin\varphi + H_\varphi \cos\varphi \qquad\qquad (I.12b)$$
$$H_z = H_z \qquad\qquad (I.12c)$$

Gradient:

$$\nabla\Phi = \hat\rho\,\frac{\partial\Phi}{\partial\rho} + \hat\varphi\,\frac{1}{\rho}\frac{\partial\Phi}{\partial\varphi} + \hat z\,\frac{\partial\Phi}{\partial z} \qquad\qquad (I.13)$$

Divergence:

$$\nabla\cdot\vec H = \frac{1}{\rho}\frac{\partial}{\partial\rho}\left(\rho H_\rho\right) + \frac{1}{\rho}\frac{\partial H_\varphi}{\partial\varphi} + \frac{\partial H_z}{\partial z} \qquad\qquad (I.14)$$

Laplacian

$$\nabla^2\Phi = \frac{1}{\rho}\frac{\partial}{\partial\rho}\left(\rho\frac{\partial\Phi}{\partial\rho}\right) + \frac{1}{\rho^2}\frac{\partial^2\Phi}{\partial\varphi^2} + \frac{\partial^2\Phi}{\partial z^2} \qquad\qquad (I.15)$$

Curl

$$\nabla\times\vec H = \hat\rho\left(\frac{1}{\rho}\frac{\partial H_z}{\partial\varphi} - \frac{\partial H_\varphi}{\partial z}\right) + \hat\varphi\left(\frac{\partial H_\rho}{\partial z} - \frac{\partial H_z}{\partial\rho}\right) + \hat z\left(\frac{1}{\rho}\frac{\partial\left(\rho H_\varphi\right)}{\partial\rho} - \frac{1}{\rho}\frac{\partial H_\rho}{\partial\varphi}\right) \qquad (I.16)$$

Spherical

The spherical coordinate system has been drawn in Fig. I.3. The coordinate and vector transformations between the rectangular and spherical systems, and the metrics, gradient, divergence, Laplacian, and curl operators are given by

Coordinates:

$$(r,\theta,\varphi) \quad 0 \le r < \infty, \quad 0 \le \theta \le \pi, \quad 0 \le \varphi \le 2\pi \qquad\qquad (I.17)$$

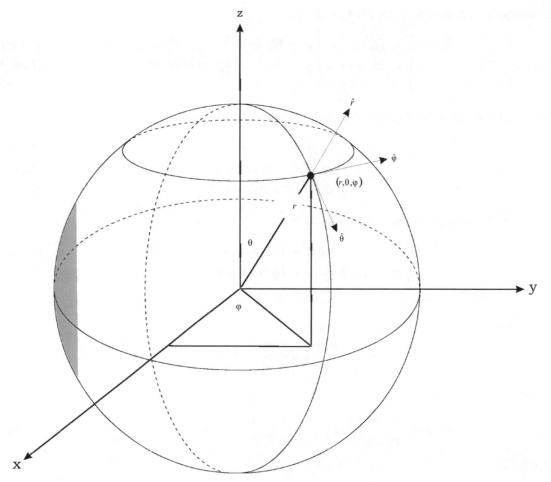

FIGURE I.3: Spherical coordinate system.

Metrics:

$$h_r = 1, \quad h_\theta = r, \quad h_\varphi = r \sin \theta \qquad (I.18)$$

Rectangular to Spherical Coordinate Transformations:

$$r = \left(x^2 + y^2 + z^2\right)^{\frac{1}{2}} \qquad (I.19a)$$

$$\theta = \arccos\left(\frac{z}{\left(x^2 + y^2 + z^2\right)^{\frac{1}{2}}}\right) \qquad (I.19b)$$

$$\varphi = \arctan\left(\frac{y}{x}\right) \qquad (I.19c)$$

Rectangular to Spherical Vector Transformations:

$$H_r = H_x \sin\theta \cos\varphi + H_y \sin\theta \sin\varphi + H_z \cos\theta \qquad (I.20a)$$
$$H_\theta = H_x \cos\theta \cos\varphi + H_y \cos\theta \sin\varphi - H_z \sin\theta \qquad (I.20b)$$
$$H_\varphi = -H_x \sin\varphi + H_y \cos\theta \qquad (I.20c)$$

Spherical to Rectangular Coordinate Transformations:

$$x = r \sin\theta \cos\varphi \qquad (I.21a)$$
$$y = r \sin\theta \sin\varphi \qquad (I.21b)$$
$$z = r \cos\theta \qquad (I.21c)$$

Spherical to Rectangular Vector Transformations:

$$H_x = H_r \sin\theta \cos\varphi + H_\theta \cos\theta \cos\varphi - H_\varphi \sin\varphi \qquad (I.22a)$$
$$H_y = H_r \sin\theta \sin\varphi + H_\theta \cos\theta \sin\varphi + H_\varphi \cos\varphi \qquad (I.22b)$$
$$H_z = H_r \cos\theta - H_\theta \sin\theta \qquad (I.22c)$$

Gradient:

$$\nabla\Phi = \hat{r}\frac{\partial\Phi}{\partial r} + \hat{\theta}\frac{1}{r}\frac{\partial\Phi}{\partial\theta} + \hat{\varphi}\frac{1}{r\sin\theta}\frac{\partial\Phi}{\partial\varphi} \qquad (I.23)$$

Divergence:

$$\nabla\cdot\vec{H} = \frac{1}{r^2}\frac{\partial}{\partial r}\left(r^2 H_r\right) + \frac{1}{r\sin\theta}\frac{\partial}{\partial\theta}\left(\sin\theta\, H_\theta\right) + \frac{1}{r\sin\theta}\frac{\partial H_\varphi}{\partial\varphi} \qquad (I.24)$$

Laplacian

$$\nabla^2\Phi = \frac{1}{r^2}\frac{\partial}{\partial r}\left(r^2\frac{\partial\Phi}{\partial r}\right) + \frac{1}{r^2\sin\theta}\frac{\partial}{\partial\theta}\left(\sin\theta\frac{\partial\Phi}{\partial\theta}\right) + \frac{1}{r^2\sin^2\theta}\frac{\partial^2\Phi}{\partial\varphi^2} \qquad (I.25)$$

Curl

$$\nabla\times\vec{H} = \frac{\hat{r}}{r\sin\theta}\left(\frac{\partial}{\partial\theta}\left(H_\varphi\sin\theta\right) - \frac{\partial H_\theta}{\partial\varphi}\right) + \frac{\hat{\theta}}{r}\left(\frac{1}{\sin\theta}\frac{\partial H_r}{\partial\varphi} - \frac{\partial}{\partial r}\left(r H_\varphi\right)\right)$$
$$+ \frac{\hat{\varphi}}{r}\left(\frac{\partial}{\partial r}\left(r H_\theta\right) - \frac{\partial H_r}{\partial\theta}\right) \qquad (I.26)$$

Prolate Spheroidal

The prolate spheroidal coordinate system has already been drawn in Fig. 2.1. The coordinate and vector transformations between the rectangular and prolate spheroidal systems, and the metrics, gradient, divergence, Laplacian, and curl operators are given by

Coordinates:

$$(\xi, \eta, \varphi) \quad 1 \leq \xi \leq \infty, \quad -1 \leq \eta \leq 1, \quad 0 \leq \varphi \leq 2\pi \tag{I.27}$$

Metrics:

$$h_\xi = c \sqrt{\frac{\xi^2 - \eta^2}{\xi^2 - 1}} \tag{I.28a}$$

$$h_\eta = c \sqrt{\frac{\xi^2 - \eta^2}{1 - \eta^2}} \tag{I.28b}$$

$$h_\varphi = c \sqrt{(\xi^2 - 1)(1 - \eta^2)} \tag{I.28c}$$

where $c = \sqrt{a^2 - b^2}$ is half the focal length of the prolate spheroid, while a and b are the half-length and half-width, respectively, of any spheroid with $\pm c$ as its foci.

Rectangular to Prolate Spheroidal Coordinate Transformations:

$$\xi = \frac{r_2 + r_1}{2c} \tag{I.29a}$$

$$\eta = \frac{r_2 - r_1}{2c} \tag{I.29b}$$

$$\varphi = \tan^{-1}\left(\frac{y}{x}\right) \tag{I.29c}$$

where

$$r_1 = \sqrt{x^2 + y^2 + (z - c)^2}$$

$$r_2 = \sqrt{x^2 + y^2 + (z + c)^2}$$

Rectangular to Prolate Spheroidal Vector Transformations:

$$H_\xi = \xi \sqrt{\frac{1 - \eta^2}{\xi^2 - \eta^2}} \cos\varphi \, H_x + \xi \sqrt{\frac{1 - \eta^2}{\xi^2 - \eta^2}} \sin\varphi \, H_y + \eta \sqrt{\frac{\xi^2 - 1}{\xi^2 - \eta^2}} \, H_z$$

$$H_\eta = -\eta \sqrt{\frac{\xi^2 - 1}{\xi^2 - \eta^2}} \cos\varphi \, H_x - \eta \sqrt{\frac{\xi^2 - 1}{\xi^2 - \eta^2}} \sin\varphi \, H_y + \xi \sqrt{\frac{1 - \eta^2}{\xi^2 - \eta^2}} \, H_z \tag{I.30}$$

$$H_\varphi = -\sin\varphi \, H_x + \cos\varphi \, H_y$$

Prolate Spheroidal to Rectangular Coordinate Transformations:

$$x = c\sqrt{\left(\xi^2 - 1\right)\left(1 - \eta^2\right)}\cos\varphi \qquad (I.31a)$$

$$y = c\sqrt{\left(\xi^2 - 1\right)\left(1 - \eta^2\right)}\sin\varphi \qquad (I.31b)$$

$$z = c\,\xi\,\eta \qquad (I.31c)$$

Prolate Spheroidal to Rectangular Vector Transformations:

$$H_x = \xi\sqrt{\frac{1 - \eta^2}{\xi^2 - \eta^2}}\cos\varphi\, H_\xi - \eta\sqrt{\frac{\xi^2 - 1}{\xi^2 - \eta^2}}\cos\varphi\, H_\eta - \sin\varphi\, H_\varphi \qquad (I.32a)$$

$$H_y = \xi\sqrt{\frac{1 - \eta^2}{\xi^2 - \eta^2}}\sin\varphi\, H_\xi - \eta\sqrt{\frac{\xi^2 - 1}{\xi^2 - \eta^2}}\sin\varphi\, H_\eta + \cos\varphi\, H_\varphi \qquad (I.32b)$$

$$H_z = \eta\sqrt{\frac{\xi^2 - 1}{\xi^2 - \eta^2}}\, H_\xi + \xi\sqrt{\frac{1 - \eta^2}{\xi^2 - \eta^2}}\, H_\eta \qquad (I.32c)$$

Gradient:

$$\nabla\Phi = \hat{\xi}\frac{1}{c}\sqrt{\frac{\xi^2 - 1}{\xi^2 - \eta^2}}\frac{\partial\Phi}{\partial\xi} + \hat{\eta}\frac{1}{c}\sqrt{\frac{1 - \eta^2}{\xi^2 - \eta^2}}\frac{\partial\Phi}{\partial\eta} + \hat{\varphi}\frac{1}{c}\frac{1}{\sqrt{\xi^2 - \eta^2}}\frac{\partial\Phi}{\partial\varphi} \qquad (I.33)$$

Divergence:

$$\nabla\cdot\vec{H} = \frac{1}{c\left(\xi^2 - \eta^2\right)}\left[\frac{\partial}{\partial\xi}\left(H_\xi\sqrt{\left(\xi^2 - \eta^2\right)\left(\xi^2 - 1\right)}\right) + \frac{\partial}{\partial\eta}\left(H_\eta\sqrt{\left(\xi^2 - \eta^2\right)\left(\xi^2 - 1\right)}\right)\right.$$

$$\left. + \frac{\partial}{\partial\varphi}\left(H_\varphi\frac{\xi^2 - \eta^2}{\sqrt{\left(\xi^2 - \eta^2\right)\left(\xi^2 - 1\right)}}\right)\right] \qquad (I.34)$$

Laplacian

$$\nabla^2\Phi = \frac{1}{c^2\left(\xi^2 - \eta^2\right)}\left[\frac{\partial}{\partial\xi}\left(\left(\xi^2 - 1\right)\frac{\partial\Phi}{\xi}\right) + \frac{\partial}{\partial\eta}\left(\left(1 - \eta^2\right)\frac{\partial\Phi}{\partial\eta}\right)\right.$$

$$\left. + \frac{\xi^2 - \eta^2}{\left(\xi^2 - 1\right)\left(1 - \eta^2\right)}\frac{\partial^2\Phi}{\partial\varphi^2}\right] \qquad (I.35)$$

Curl

$$\nabla \times \vec{H} = \frac{\hat{e}_\xi}{c} \left[\frac{1}{\sqrt{\xi^2 - \eta^2}} \frac{\partial}{\partial \eta} \left(\sqrt{1 - \eta^2} \, H_\varphi \right) - \frac{1}{\sqrt{(\xi^2 - 1)(1 - \eta^2)}} \frac{\partial H_\eta}{\partial \varphi} \right]$$

$$+ \frac{\hat{e}_\eta}{c} \left[\frac{1}{\sqrt{(\xi^2 - 1)(1 - \eta^2)}} \frac{\partial H_\xi}{\partial \varphi} - \frac{1}{\sqrt{\xi^2 - \eta^2}} \frac{\partial}{\partial \xi} \left(\sqrt{\xi^2 - 1} \, H_\varphi \right) \right]$$

$$+ \frac{\hat{e}_\varphi}{c} \left[\sqrt{\frac{\xi^2 - 1}{\xi^2 - \eta^2}} \frac{\partial}{\partial \xi} \left(\sqrt{\xi^2 - \eta^2} \, H_\eta \right) - \sqrt{\frac{1 - \eta^2}{\xi^2 - \eta^2}} \frac{\partial}{\partial \eta} \left(\sqrt{\xi^2 - \eta^2} \, H_\xi \right) \right] \qquad (I.36)$$

A complete reference for all vector operators useful in electromagnetic modeling can be found in [1], along with an in-depth mathematical treatment of curvilinear coordinate systems. The prolate spheroidal system is treated more thoroughly in [2] and [3].

REFERENCES

[1] G. B. Arfkin and H. J. Weber, *Mathematical Methods for Physicists*. San Diego, CA: Academic, 2001, pp. 103–131.

[2] G. B. Arfkin, *Mathematical Methods for Physicists*, 2nd ed. New York: Academic, 1970, pp. 72–107.

[3] P. M. Morse and H. Feshbach, *Methods of Theoretical Physics*, Part I. New York: McGraw-Hill, 1953.

Appendix II

BIOT–SAVART LAW

The Biot–Savart law can be derived from the relationship between the magnetic vector potential $\vec{A}$ and current density $\vec{J}$. Writing (3.10) in the Cartesian coordinate system gives

$$\vec{A}(x, y, z) = \frac{\mu_0}{4\pi} \int_{V'} \frac{\vec{J}(x', y', z')}{\sqrt{(x - x')^2 + (y - y')^2 + (z - z')^2}} dv' \qquad (II.1)$$

where the coordinate system is defined in Fig. I.1. If a line of constant current I flows along the z-axis from a to b, then $\vec{J}(x', y', z') \, dv' = I \, dz' \, \hat{a}_z$ and (II.1) becomes

$$A_z(x, y, z) = \frac{\mu_0 I}{4\pi} \int_a^b \frac{1}{\sqrt{x^2 + y^2 + (z - z')^2}} dz' \qquad (II.2)$$

Computing $\nabla \times \vec{A}$ results in

$$B_x(x, y, z) = \frac{-\mu_0 I y}{4\pi} \int_a^b \frac{1}{\left((x - x')^2 + (y - y')^2 + (z - z')^2\right)^{\frac{3}{2}}} dz' \qquad (II.3)$$

$$B_y(x, y, z) = \frac{\mu_0 I x}{4\pi} \int_a^b \frac{1}{\left((x - x')^2 + (y - y')^2 + (z - z')^2\right)^{\frac{3}{2}}} dz' \qquad (II.4)$$

where B_x and B_y are the components of magnetic flux density in Cartesian coordinates. Integrating (II.3) and (II.4) and simplifying gives

$$B_x(x, y, z) = \frac{-\mu_0 I y}{4\pi (x^2 + y^2)} \left(\frac{z - a}{[x^2 + y^2 + (z - a)^2]^{\frac{1}{2}}} - \frac{z - b}{[x^2 + y^2 + (z - b)^2]^{\frac{1}{2}}} \right) \qquad (II.5)$$

$$B_y(x, y, z) = \frac{\mu_0 I x}{4\pi (x^2 + y^2)} \left(\frac{z - a}{[x^2 + y^2 + (z - a)^2]^{\frac{1}{2}}} - \frac{z - b}{[x^2 + y^2 + (z - b)^2]^{\frac{1}{2}}} \right). \qquad (II.6)$$

Equations of the form of (II.5) and (II.6) can be used to approximate the magnetic fields from air-core degaussing coils using straight line segments to represent the cable configurations of loops with complex shapes. However, care must be taken when modifying (II.5) and (II.6) for current segments lying in directions other than along the z-axis.

Author Biography

John J. Holmes (S'73–M'77–SM'00) received the B.S., M.S., and Ph.D. degrees in electrical engineering from West Virginia University, Morgantown, in 1973, 1974, and 1977, respectively. He joined the Naval Surface Warfare Center in 1977 and is currently the Senior Scientist for the Underwater Electromagnetic Signatures and Technology Division, where he is engaged in research on underwater electromagnetic field signature reduction systems for surface ships and submarines. He is the author or coauthor of two books and 25 peer-reviewed papers. He is the holder of ten patents.

Dr. Holmes was the recipient of the Meritorious Civilian Service Award in 1986 and the David Packard Excellence in Acquisition Award in 1999.

Printed in the United States
by Baker & Taylor Publisher Services